Seeking Communion as Healing Dialogue

Seeking Communion as Healing Dialogue

Gabriel Marcel's Philosophy for Today

Margaret M. Mullan

LEXINGTON BOOKS
Lanham • Boulder • New York • London

Published by Lexington Books
An imprint of The Rowman & Littlefield Publishing Group, Inc.
4501 Forbes Boulevard, Suite 200, Lanham, Maryland 20706
www.rowman.com

6 Tinworth Street, London SE11 5AL, United Kingdom

British Library Cataloguing in Publication Information Available

Library of Congress Cataloging-in-Publication Data

Name: Mullan, Margaret M., 1976–, author.
Title: Seeking communion as healing dialogue : Gabriel Marcel's philosophy for today / Margaret M. Mullan.
Description: Lanham : Lexington Books, [2021] | Includes bibliographical references and index. | Summary: "This book explores society's problems with interpersonal communication amid increasingly technological environments. The author argues that the work of Gabriel Marcel reveals the root of our issues with communication to be issues with being with others, ultimately suggesting that seeking communion is a way to bridge our disconnections"—Provided by publisher.
Identifiers: LCCN 2021006383 (print) | LCCN 2021006384 (ebook) | ISBN 9781793621771 (cloth) | ISBN 9781793621788 (ebook)
Subjects: LCSH: Marcel, Gabriel, 1889–1973 | Interpersonal communication—Technological innovations.
Classification: LCC B2430.M254 M85 2021 (print) | LCC B2430.M254 (ebook) | DDC 153.6—dc23
LC record available at https://lccn.loc.gov/2021006383
LC ebook record available at https://lccn.loc.gov/2021006384

"Upon this gifted age in its dark hour, rains from the sky a meteor shower of facts . . . they lie unquestioned, uncombined. Wisdom enough to leech us of our ill is daily spun, but there exists no loom to weave it into fabric."
—Edna St. Vincent Millay

Encountering Gabriel Marcel's insights on living in this age, in what Marcel would call "the Broken World," I saw the vague outline of a loom for weaving meaning from these experiences. Gabriel Marcel was a person who stopped to explain how he made sense of living amid a world of uncertainty. What one person identifies as a loom for making sense of life may not be the loom used by next person. This book is dedicated to all those who pause to think about the loom, the meaning to be woven from piles of facts and experiences that so often lie "unquestioned, uncombined." To all those who pause, thank you for taking the time to put into words, spoken or written, what gives meaning to our interactions during our own "gifted age."

Contents

Acknowledgments

Writing a book on any topic, even on the subject of dialogue, *seems* to be a mostly solitary experience. The bulk of the writing of this project, was carried out in an even more markedly solitary year—2020. However, contrary to appearances, the book-writing process expresses an ongoing dialogue with other voices still living or in texts. Typing quietly in my home office with the sound of a small brook streaming past my window *actually* means I am engaging in dialogue with past authors and future readers. This project began with in-person conversations, continued in solitary research, transformed during more in-person discussions, and hopefully continues growing as dialogue with persons who are both readers and in-person conversants.

I am filled with gratitude to all who dialogued with me along this journey. Michael Hyde states that "acknowledgment is a life-giving gift"[1] and I would add, acknowledging the givers is also a life-giving act.

This book expresses ideas in response to dialogues initiated during my doctoral program in rhetoric at Duquesne University. The scholarship about the philosophy of communication shared in classes and discussions by my professors Ron Arnett, Pat Arneson, Garnet Butchart, Erik Garrett, and Janie Harden-Fritz inspired my exploration of the meaning of dialogue today. Pat Arneson, thank you for guiding me through the dissertation process, our conversations opened doors for my own discovery and ambitious deadlines moved me to start and stay in action to complete my dissertation. Thank you to all my Duquesne colleagues, especially Jenn Lo Castro, Robert Foschia, and Tim Michaels, who continue to offer insight about the process of sharing ideas through scholarship.

Although this book bears a general resemblance to the themes of my dissertation—a study of Gabriel Marcel's insights on intersubjectivity and technology—I have rewritten all six chapters, extracting more connections between Marcel's philosophy and our problems today. The task of rewriting this dissertation has unfolded in fits and starts. Starting a tenure-track position at East Stroudsburg University in Fall 2017 meant I had little time to address this project.

I am deeply grateful to my colleagues at East Stroudsburg University for inviting me to join this exceptional department—outstanding in its collegiality and authentic support of me as teacher, scholar, and friend. I also thank you for your support of my scholarship by offering to help me navigate teaching demands whenever possible. Thank you, Andi McLanahan, Rob McKenzie, Wenjie Yan, Cem Zeytinoglu, Paul Lippert, and Charlie Warner for your ongoing support for me as colleagues and friends—each of you always willing to dialogue about teaching, writing, and/or life.

I am most grateful to family and friends who have encouraged and supported my starting and stopping along the way. I am grateful to my parents, Dermott and Susan Mullan, who are forever encouraging of my process, of my work, of my taking breaks at their home. I count our friendship as real experiences of communion and dialogue. I am grateful for each and all of my siblings, especially during 2020, our ongoing virtual meetings that always involved them checking in with me about how I was handling writing in 2020. I thank all my siblings and their spouses for their encouraging presence and advice along the way. And I thank my friends who accompanied me in increasing freedom from earlier journeys.

I am grateful to teacher-scholars I met in my master's program: among them, Robert H. Woods Jr., who introduced me to dialogue and media ecology scholars, the Duquesne University program, and the idea that I could find fulfillment as a professor. A dialogue with him empowered me to envision my life as completely other to what it was at that time. Maybe that was an experience of what Gabriel Marcel calls "*syneidesis* . . . a kind of vision which brings things together," a *syneidesis* which uncovers new beginnings. I do not know if this work will be my only book, but I do know it is my first, a beginning of a dialogue. All the persons, the conversations, and the support combined with my starting and restarting to write have brought me to this moment: sharing with you my thoughts on dialogue for today.

NOTE

1. Hyde, M. J. (2004). "The Ontological workings of dialogue and acknowledgment." In R. Anderson, L. A. Baxter, & K. N. Cissna (Eds.). *Dialogue: Theorizing difference in communication studies* (57–74). Thousand Oaks, CA: SAGE.

ONE

Dialoguing on Edge

People of every age gaze out at their world and feel that living itself is falling apart,[1] has been fractured, and threatens to continue disintegrating throughout. I write this from within the global collapse brought on by COVID-19 in the year 2020. I write this during the local, national, and global cries for justice for black persons in the U.S. The killing of George Floyd on May 25, 2020, brings to national attention racism embedded and embodied in our U.S. criminal justice system.[2] This moment of once-again exposing the systemic brokenness in treatment of black communities in the U.S. may be a watershed moment for addressing and reforming attitudes, actions, and systems.[3]

Our world feels seismically broken along a myriad of fault lines. I stand at the edge of just one of those rifts: our communication. On a global scale, international and intercultural communication about vital information, research, and response is deeply broken in ways that have and will continue to bear devastating consequences for the present and future people in our world. On a community level, interpersonal communication about what is happening in our communities is also deeply broken. Persons speak about what they think or feel should be a response and others hear their thoughts and feelings and cannot even understand what they mean. Persons share and comment on "news" about global, national, or local response. While they use the same language to talk with others, participants in many conversations sense that they have no idea what the other person means. At the most fundamental level of communication, person-to-person conversation, we find a lack of understanding and a breakdown in communication. We stand across from the other person, on opposite sides of a chasm, with no visible way across to meet the other person.

In the best of times, our interpersonal communication, as a process of sharing meaning, is fraught with barriers, threats, uncertainties, limitations, and impossibilities. Not all (nor many) conversations between persons (again, in the best of times) become genuine dialogue: an encounter between persons in the context of a relationship during which both persons share an understanding. Genuine dialogue involves what philosopher Martin Buber described as, "turning toward the other."[4] Turning toward the other, we somehow cross this chasm between us and meet the other person face-to-face or person-to-person. Dialogue is more than conversation because dialogue means more than transmission of messages.[5] In moments of being present with and to each other, we know what the other person means. In dialogue, we are really *with* the other person and we understand the other person who is *with* us. Buber's description of dialogue suggests an "authenticity of being" in this really being *with* the other.[6] Dialogue spans fractures of misunderstandings, disconnects, or dismissals. Genuine dialogue sometimes emerges in the exchanges between persons and *many times* remains elusive.

Communicators seeking to understand and be understood in face-to-face conversations encounter multiple challenges to experiencing genuine dialogue. A communicator seeking to express an idea or feeling meets potential obstacles at every stage in the process. I attempt to translate my thoughts or feelings into words as I struggle to give form to what I *mean to* say. The expressed words are voiced into a physical or a virtual context constructed of currently emitting or echoes of previously emitted sounds. Will my words sound through the existing noise? Will the receiver hear the words at all? If they hear my words, will they understand what I meant to say? In a split-second, my desire to speak my mind so as to be understood faces an endless stream of actual or potential, physical or emotional uncertainties threatening the entire process of genuine communication.

Along with the stream of uncertainties about understanding and being understood, technology brings added dimensions to our search for genuine dialogue. Our face-to-face conversations may happen in-person in actual[7] shared space and time and/or in virtual shared online spaces and times. While many of us are experts in—or at least used to—communicating with and amid technologically mediated settings, our technologies bring extra—sometimes complicated—dimensions to our conversations. We hear that our technology promises to *improve* our communication and strengthen our connections[8] and we see evidence that sometimes we share more about ourselves via technology in "hyperpersonal" communication.[9] Technology opens up real possibilities for meaningful virtual encounters and enhanced actual encounters.

Communicating with or amid technology may also bring more uncertainties for the participants seeking dialogue. While speaking via technology, more questions emerge about our communication: Can they hear me? Can they see me? Can they sense I am really here? Are they distracted? Am I distracted? Is this person the actual person I think I am talking to? Is it just them or are there other people also present? These uncertainties build alongside the already existing questions about understanding.

In an acknowledgment of my biases, I believe these potential or actual barriers created by technology are not the source of our problems in experiencing genuine dialogue. Standing on either side of our chasms of misunderstanding, we as human communicators may be helped by technology as mediator and as way-for-communication or may be hindered by technology as barrier to and gap in communication. We, the human communicators, are *the source of our struggles* in turning toward the other, understanding the other, connecting with the other, and being *with* the other. We, the human communicators, misread the tone in a text message. We, the human communicators, jump quickly to judgment of fellow communicators based on what we heard them say. We, the human communicators, set up walls to block and cancel the sound of the other's presence online or on our mobile devices.

Although I believe human communicators are the source of our miscommunications and misunderstandings, our technologies have changed the nature and dimensions of these interpersonal misunderstandings. Because our messages sent via technology are permanent (recorded and undeletable), the sound or images of the messages remain echoing and stamped on screens, paper, or as computer memory cells. A primary principle of communication, one can never unsay what one has said, assumes a more permanent presence in the form of digital messages.[10] Knowing our texts or images may be permanently present may bring us fear that we will lose control of how that text or image is understood. Our message sent without much forethought reverberates forever. Our messages are sent as bits per second and the speed with which the message is sent is always getting faster with new developments. In a matter of seconds, a message meant for one person may be posted online and read by hundreds of thousands. We may be communicating with an exponentially increasing number of people, known to or unknown to us. Our messages are separated from the context in which they were sent. Because the message is dislocated from context, the meaning may be more easily misunderstood. Technology changes the nature of our communication in time and space, decontextualizing our messages and changing what our messages mean. Technology changes the way we stand along these chasms between us.

Our interpersonal conversations are fraught with obstacles to genuine dialogue, to really being present to one another. We speak our minds and feel that we are not heard. We hear others communicating with us, but we are not really listening. We are with other people who are speaking, texting, or typing messages to us, but we feel like we are not really with them. We try to use words to build bridges between us where we can meet to express what we think, feel, or want. Yet sometimes those bridges run parallel to each other or fall short of actually meeting in the middle, and then I am left on one side of a chasm, while the other is over-there, beyond earshot and/or out-of-sight, beyond communication. Where do we start to find a solution for these problems? Can we perfect our selection of words? Can we build our bridges between us better? Are we headed off in wrong direction with the bridge? Are we standing at the edge of interpersonal chasms that simply cannot be spanned?

In my search to understand this problem of communication, I stumbled upon this description. The author described an experience between two persons who are standing in the same room. Between the two persons,

> Merely physical, communication is possible; the image of passing of messages between a reception point and an emission point. Yet something essential is lacking. One might say that what we have with this person, who is in the room, but somehow not really present to us, is communication without communion: unreal communication, in a word. He[11] understands what I say to him, but he does not understand *me*: I may even have the extremely disagreeable feeling that my own words, as he repeats them to me, as he reflects them back to me, have become unrecognizable. By a very singular phenomenon indeed, this stranger interposes himself between me and my own reality, he makes me in some sense also a stranger to myself; I am not really myself while I am with him.[12]

This passage by a twentieth-century French philosopher, Gabriel Marcel (1889–1973), described the same problems in communication I was trying to understand. While the message being transmitted is understood, the person who delivers the message does not feel understood. The "merely physical communication" may be happening, but *something more* is missing. We can (and do) share a physical space with someone and feel they are absent from us. "He understands what I say to him, but he does not understand *me*."[13] I seek to share not just my messages but who I am. When the other person does not also understand *me*, the one who communicates, I feel as if our communication has failed. Feeling one's own words are now unrecognizable, one feels alienated from one's own words uttered moments earlier. Communication, meant to bridge a gap between persons, has left me feeling more isolated from the other person and even alienated from my own words.

Genuine dialogue today—bridging our gaps between us—needs more than transmission of messages. A communicator may pass along verbal or nonverbal messages to the other person and still not be in dialogue with her. Two persons may share the same language, the same channels for communicating, and yet feel they are not understood. People may share physical or virtual spaces and they may sense the other person is *there* in the same space and still feel the other is not present with them. Even as technology gives us bridges across physical distances, these bridges may not always bring us closer to the other person. Today, as always, two communicators need more than communication as grounds for the birth of genuine dialogue.

Marcel points to what that "more than communication" might mean. Marcel followed the scenario detailing a lack of communion with a distinct experience. The two persons are both in the same room, "When somebody's presence does really make itself felt, it can refresh my inner being; it reveals me to myself, it makes me more fully myself than I should be if I were not exposed to its impact."[14] Marcel put into words an interpersonal experience of being *with* the other. Marcel's description of presence as "presence making itself felt" and refreshing one's inner being sheds light on communicative experiences. Communication as being *with* the other involves more than messages transmitted; it is a deeper sharing of being. When participants sense the presence of the other person, they not only share ideas, but also share something more. This project explores what Marcel's phrase "more than communication" means for communicators today.

GABRIEL MARCEL'S APPROACH TO BEING IN A BROKEN WORLD

Reading the phrase "communication without communion: unreal communication," I was intrigued by what Marcel meant about our problems in communication. Beginning with *Mystery of Being*, which contained this cited reflection on communion, being, and presence, I dove into his books and plays. In 1914, Marcel began to record his reflections in his philosophical diary *Metaphysical Journal*, and he wrote, spoke, and discussed his ideas about being in the world until his death in 1973.[15] I started chronologically with Marcel's reflections in his *Metaphysical Journal*, first started in 1914 and published in 1929. Then I moved through his essays and books: "Ontological Mystery" in 1933, *Being and Having* in 1935, *Creative Fidelity* 1940, *Homo Viator* in 1945, and *Man Against Mass Society* in 1951, through to *Tragic Wisdom and Beyond* in 1971.[16] Delving into Marcel's ideas about communication and being *with*, I encountered several important authors whose works introduced me to Marcel's wide range of philosophical reflections, plays, and insights.[17]

Marcel's philosophy opens up the communication problem of bridging gaps between people in genuine dialogue. The gaps of broken communication appear because our world itself is broken. Identified by Jean Wahl as a philosopher of existence,[18] Marcel attends to concrete experiences and attitudes lived in "the broken world."[19] Marcel uses the word "world" for a broad, changing context and for a personal living experience. Marcel studied the world as he experienced it in Western European France from the 1910s through the 1960s. He reflected on the situation of his Parisian society, of his French setting, of living through two World War and being afraid of nuclear war during the 1950s and 1960s. He made macroanalyses of the effects of technological developments he witnessed unfolding in France throughout his life. Marcel brings these reflections from macro-contexts to one's personal world. Each person is a "being-in-a-situation" dwelling in a particular set of circumstances.[20]

Marcel's interpretation of the world was permeated by his attention to *being* and a concrete, or lived, experience. His earliest writings—recorded between 1910[21] and 1914 in his *Metaphysical Journal*—reveal reflections on what being is and how presence is connected to it. In his initial years of philosophical reflections, Marcel assumed an idealist approach, a method of privileging abstract ideas as the way to understand the truth about reality.[22] In his early writings, he used abstract terms to describe how he viewed being, knowledge, and life.[23] Marcel's experience working for the French Red Cross in World War I, called his attention to concrete, lived experiences. He worked in the office of missing persons, where people arrived daily, anguished by uncertain loss, and asking for help to locate missing loved ones. During and after his experience in World War I, Marcel turned from writing about the *idea* of existence to exploring the meaning of *lived* drama of existence.[24] The concrete reality of this person suffering before Marcel, the subject, acted upon him in ways he could not ignore. Marcel called this "concrete approach" a study of "life as it is concretely lived."[25] In his concrete approach, one begins by taking actual experiences and situations in life and then proceeds by reflecting on what these experiences might mean or uncover, "so that [one] may try to throw more light upon life."[26] Marcel's study of being involves reflecting on actual life experiences and then stepping back to understand what these experiences might mean.

One can only study being as a particular being in a situation. For example, in a society privileging technological efficiency, one's life is evaluated according to one's work productivity.[27] A human person feels reduced to the rate at which they produce or work. Marcel calls this a problem of the technical world: in our commitments to technological efficiency, we overlook the changes experienced by persons living in this world. Marcel names this common experience a characteristic of living in "the broken world."[28] Our contemporary experience of the broken world cannot be reduced to the exact characteristics Marcel uses to describe life in his Western European world in 1930s, 1940s, or 1950s. And yet, on some level, we share similar problems with our world's former inhabitants who shared our same human condition.

"The broken world," as a category[29] of human experience, bears external, public societal dimensions and internal, personal dimensions. *Inter*personal communication begins as an act of *intra*personal communication: becoming aware of who I am and being able to put my thoughts into words. The way one sees and talks to oneself blends with one's feelings and awareness about oneself. Problems in interpersonal communication may begin as problems in *intra*personal communication. For different reasons, today's communicator does not often reflect on "who I am," the one who is in communication with others. Today's communicators do not frequently reflect on being in the world because they do not know *how to* reflect about who they are in the world. People in the broken world do not understand what *being* in the world means.

Marcel connects this neglect of one's own sense of being in the world to a devaluation of being human. Reductive views of the human person strip the human person of a belief that her individual life *is* sacred. In the broken world, people have lost the sense of what it means to be a human being. Marcel sees that because one *is*, one's life is "sacred."[30] Marcel's insights into sacredness of living emerges within a philosophical and ethical landscape of scholarship on how life, lives, and ecologies bear sacred dimensions. [31]

In his approach to philosophy, Marcel sometimes uses spiritual or religious dimensions to describe human experience. Because of his study of mystery, being, and transcendence, Marcel is often named as a Christian existentialist—what he called a misnomer for the whole of his philosophical work.[32] Marcel did not see himself as first a theologian seeking philosophy (existential or otherwise) for explanations. In fact, Marcel began his philosophical writing in 1913 while not affiliating with any religion and only recorded a personal spiritual experience in 1929.[33] He preferred being identified as a wayfaring philosopher[34] who reflected on what it means to be human.[35] The study of being human may sometimes include experiences of phenomena beyond the philosophical or scientific explanations. Marcel noted how people who objected to his philosophy about *being* in the world often asked him,

> But what can be the value of such a philosophy for those who are a-Christian—for those who ignore Christianity or do not accept it? I would answer: it is quite possible that the existence of the fundamental Christian data may be necessary *in fact* to enable the mind to conceive some of the notions which I attempted to analyze; but these notions cannot be said to depend on the data of Christianity, and *they do not presuppose it*.[36]

Philosophers, who in their personal lives espouse religious belief, may still offer valid insights into a shared human experience that transcends or underlies one's individual interpretations of our world.[37]

In his description of a human life as being sacred, Marcel states that if one does not believe that this human life reflects a divine image, one still must acknowledge a "mystery inherent in the human condition," a sacredness in this human life that must be respected because she/he/they is.[38] The sacredness of a life comes into relief when people treat the life of the other as if it were worth nothing, degrading them and killing them in concentration camps or in war.[39] To deny that the life of this person is always sacred and thus inviolable is to open the door to "radical destruction" of all that we call being human.[40] Marcel's philosophy assumes this as his starting point: each and every human being is sacred because she/he/they *is*.

When we step away from this certainty that this human person is sacred, we assign less value to this human person. Ignoring who the human person is—as more than a name, an abstraction, a producer, a part in the machine—leads to degradation in our communal world and in intrapersonal worlds. Marcel identifies this ignorance of who the human person *is* as one of the greatest problems in today's world. Uncertain of her own value, a communicator in a technological society struggles to understand who she is in this world.

PROBLEMS WITH TECHNOLOGY, PROBLEMS WITH BEING, PROBLEMS WITH COMMUNICATION

Marcel believed philosophy should only be engaged in response to a personal question, a personal need, a personal desire to understand. This project springs from my desire to understand why we are so often unable to communicate with the people at our side. In studying communication, we attend to the context for, the participants, the messages, the coding, the translations, the understanding, the goals of and the (im)possibilities for the encounter. These components of communication offer multiple areas for potential glitches, problems, and failures. Seeking explanations of why we struggle to communicate with others, in this project, I explore how technology forms important context for communicators who participate in/as the problems and possibilities for interpersonal communication.

This project joins an exploration of the meaning of dialogue that unfolds within the communication discipline.[41] Many scholars have reflected upon Martin Buber's philosophy of dialogue.[42] Contemporary philosophies of dialogue within the communication discipline call on descriptions offered by a number of other twentieth-century philosophers: Mikhail Bakhtin, Martin Buber, Hans Georg Gadamer, Jurgen Habermas, Martin Heidegger, and Emmanuel Levinas.[43] Philosophy scholars have offered significant studies of Gabriel Marcel's approach to intersubjectivity.[44] In a landmark article in communication literature about dialogue, Richard Johannesen identified Gabriel Marcel as a scholar who named dialogue as key to understanding the human person.[45] Few communication scholars have explored this comment by Johannesen on Marcel's approach to dialogue.[46] This project extracts Marcel's contributions to our understanding of dialogue from and within a communication perspective.

This study of dialogue unfolds in a context changed and shaped by our use of technology. Technology has brought changes for our perceptions of, understandings of, embodied living in, and communication about our world. We voice questions about our human relationships amid, alongside, and in spite of our technology. Our contemporary experiences changed by technology invite a philosophy of existence that studies shared human condition and this particular experience.[47] Marcel scholar Thomas Anderson used Marcel's ideas about technology and being to study the possibilities for and limitations to embodied persons interacting in virtual communities.[48] This project investigates if Andersons' and other philosophers' discussion of Marcel still applies to our current experiences of technology.[49] One communication scholar, Dennis Cali who studies communication, culture, and technology, notes how Marcel's philosophy suggests "losses" in our communication when we use technology.[50] Cali ends the article suggesting that "availability and Presence create the environment that creates communion" and "an ecology of human Intersubjectivity combats the deleterious effects of technology."[51] This project extends Cali's discussion by identifying *how* being available to others creates ground for communion amid our technological environments.

In 1988, Donald F. Traub highlighted how Marcel's philosophy opens up our understanding of the effects of technological changes on human existence. Traub notes the many ways living in a broken world means alienation from others.[52] Marcel's proposal for intersubjectivity offers a building block for a "true fraternal society—a society in communion."[53] In distinction from Traub's focus on broader societal interactions, this project keeps the focus on *interpersonal* interactions: how intersubjective communication happens amid technology. Much has changed in our technology since Traub offered his rich insights in 1988. Traub ended his work, inviting thinkers from diverse professional backgrounds to heed Marcel's call "within his own proper field, pursue an unrelaxing struggle for man, for the dignity of man, against everything that today threatens to annihilate man and his dignity."[54] This work responds to Traub's invitation and returns to excavate Marcel's texts from within the field of communication.

As a communication scholar, I attend to the applications of Marcel's ideas in the understanding and practices of dialogic communication for now. I approach communication as our ways of relating to and in our worlds—communal, interpersonal, and intrapersonal worlds. This discussion joins decades-long robust conversations in diverse realms of communication scholarship about relational dialectics and dialogism,[55] communication as process and systems,[56] and the pragmatic dimensions of human communication that always already happens in relationships.[57]

This project studies, makes evident, and uncovers communication as how we understand and express our relating to our worlds.[58] This uncovering of how we communicate is situated within studies of philosophies of communication.

Today's renewed interest in Marcel's works indicates that his philosophical reflections continue to glean meaningful insights into our world.[59] As we voice questions about our human relationships amid, alongside, and in spite of our technology, philosophers like Gabriel Marcel who study existence offer us pathways into these questions.[60] Throughout his work, Gabriel Marcel speaks about struggles to be understood by the other person. In a conference organized by Martin Heidegger in Cérisy, Marcel showed marked interest in "the problem of communication."[61] Communication scholars have reflected on how philosophies open up our understanding of communication. Current edited volumes on twentieth-century philosophers who contribute to this exploration of communication or who offer philosophies of communication do not yet include significant reflection on Gabriel Marcel.[62] These explorations by communication scholars extract how philosophical discourses on, approaches to, and studies of communication contribute to our understanding of communication. While these philosophy of communication works attend to twentieth-century philosophers William James, Charles S. Peirce, Martin Buber, Martin Heidegger, Ludwig Wittgenstein, Emmanuel Levinas, Hans-Georg Gadamer, Mikhail Bakhtin, Paul Ricœur, Roman Jakobson, and Maurice Merleau-Ponty,[63] these edited works leave openings for reflecting on Gabriel Marcel's contributions to the study of communication.

The process of reading this book may lead the reader to uncover more questions than answers. My approach to studying communication involves this type of philosophy: "a philosophy is characterized more by the *formulation* of its problems than by its solution of them."[64] Philosophy is how we identify our problems and how we formulate our questions. Philosophy of communication is how we identify our problems with communication and how we formulate our questions about communication. I believe that the formulation of our questions helps us identify what may be at the root of our problems.

CHAPTER SUMMARIES

If one is hoping for practical applications of Marcel's ideas for how we communicate today, one will be disappointed by the absence of detailed applications. As an exploration of Gabriel Marcel's philosophy, this work predominantly focuses on his texts. I excavate[65] Marcel's works for reasons why today, in our technological environments, we struggle to dialogue. This excavation of Marcel's texts means gathering from across his works his approach to technology in Chapter Two, his descriptions of being in Chapter Three, his philosophy of body in Chapter Four, and his writing on intersubjectivity in Chapter Five. In this gathering of Marcel's ideas, I engage a Gadamerian art of dialogue as questioning if these texts help us understand our particular time.[66] All the ideas that I include in this book were Marcel's that still speak to us today. At the end of each chapter, I reflect on ways his themes connect to our communication. In Chapter Six, the final chapter, I offer a longer discussion of how Marcel's ideas apply to how we communicate amid technology today.

In the next section of this chapter, I identify my points of entry into Marcel's reflections. This project began by my asking, how has technology significantly impacted our contexts, channels, and ways of trying to dialogue with others? Significant impact could mean technology creates possibilities for easier or improved dialogue and significant impact could also mean technologies diminished our ability and actual dialogue.

Before attending to our communication problems, this project reflects on technology's role in our contexts for and participation in our world. Gabriel Marcel in his reflections on modern experiences of a Broken World uncovers strong connections between increased technologization and our human experiences that the world feels broken. As I culled Marcel's descriptions of technology, his questions emerged: do technics[67] promote and/or degrade human living? Chapter Two summarizes Marcel's reflections throughout his works on how technics impact our environments, our relation to ourselves and our relations with others. To understand technology's impact on our relation, I situate Marcel's approach to technics within philosophies of technology, noting how Marcel's reflections contributes to media ecology discussions and scholarship. Marcel's approach to technics contributes insightful details on our diminishing sense of the sacred in our how we view our world and cohabitants amid our technologies. Our technologies assume a vital role in how we communicate in our Broken World as context, channel, and change, and as such, Chapter Two sets the stage for the study of our problems with communication in our modern Broken World.

Next, this project turns to study the persons at the heart of our communication problems. Chapter Three begins by exploring how modern persons share feelings that life itself is broken. Existential uneasiness, "metaphysical uneasiness," is felt as being "uncertain of one's centre, it is to be in search of one's own equilibrium."[68] Marcel studies what this feeling means and reveals about a person's experience of being in the world. Our metaphysical uneasiness points to problems with being. Marcel invites learning philosophy reflection as a critical questioning of and about oneself. In this questioning, Marcel points to ontological exigence as root to and signal of our foundational problems being at all in the world. Chapter Three begins with Marcel's descriptive observations of our feelings of uneasiness, moves through *how* to engage philosophical reflection, and ends at the heart of our problems: who am I and how do I feel about my own being in the world? Chapter Three describing how one understands oneself and one's being in the world lays important ground of *intra*personal communication that will open one to *inter*personal dialogue.

Gabriel Marcel calls attention to that which is overlooked. Our increased technology use, which affords us freedom from our bodies through virtual participation, sometimes results in us forgetting to think about *what* our bodies mean in our technological worlds. What does it mean *that* I am at all? Chapter Four calls attention to our experience as bodies in our world. What does it mean *that* I am as body? Turning from one's feelings and internal world, Chapter Four unpacks *how* one participates in the world as or through internal/external participation, as body. First this chapter reflects on the meaning of my existence. Marcel exclaims "Here I am!" and then asks, what does "here I am" mean? These questions and statements may seem too theoretical to have any impact on how we communicate; however, this chapter on Marcel's philosophy of body uncovers vitally important ways that we are able to communicate at all. Because I am here and now, I am incarnate being, as body in the world, I can participate in the world and therefore can assume positions for communication. Communication scholars have explored Maurice Merleau-Ponty's insights into body as situated, as expressive and therefore as communicative.[69] This project expands upon these embodiment studies with Gabriel Marcel's description of body as positioning for communication. Many of our problems in communicating begin as questions if the other person is really here with us. What does it matter if I am aware how I am as body? As body, I am able to be in position for dialogue.

In Chapter Five, I return to our problems with communication. Many times, we stand along chasms between ourselves and others, and we ask, why do we feel so disconnected? Even when we are surrounded by people, why do we feel as if we were alone? We experience problems with sensing, understanding and acknowledging the other. We feel that the other does not see, hear, understand, or acknowledge us. In Chapter Five, I pull together Marcel's explanations throughout his works for how we sense, understand, and choose ways of being *with* each other. Marcel does not offer a comprehensive, structured explanation for intersubjectivity, but his generative insights offer attitudes and approaches that prepare/open communicators for dialogue. Because Chapter Five deals with our problems in communicating, this chapter includes multiple sections on intersubjective dimensions that Marcel notes are vital for being open for communication.

Because Marcel's starting point for all existence is life as reply, Chapter Five begins with what Marcel believes to be at the core of all relations and existence itself. Secondly and thirdly, this chapter explores how to recognize the invocations of presence and then how to select an I-thou address for the other. Marcel offers creative descriptions of intersubjective attitudes that signal attitudes toward communication. Fourth, Chapter Five summarizes these intersubjective attitudes *disponibilité* and *indisponibilité* as positions open to or closed off from others.[70] Finally, Chapter Five brings us to a question at the heart of this project: what do we mean by communion being more than communication? Marcel's description of communion uncovers meaningful ground for further exploration in various areas of communication. Marcel's insights into mid-twentieth-century communication problems continue to explain our communication problems today. In this seeking answers to our problems with communication, we also find in Marcel's philosophy, signs of what our communication could become.

In the final chapter, Chapter Six, I return to my initial question, why are we so often unable to dialogue with the people at our side? This chapter returns to study of *our* communication broken by misunderstandings, uncrossable gaps, and absence, *our* interpersonal communication complicated by the presence of our technology. I find Marcel's writing on technology, on being, and on intersubjectivity offer places to start understanding our own communication amid technology. This project, building upon connections made by other scholars,[71] uncovers rich intersections between Marcel's philosophy of technics and his insight into intersubjectivity. Marcel's insights into our problems in *really* communicating also help us understand our problems communicating amid technology.

In Chapter Six, I draw on Marcel's reflections on technology in Chapter Two to answer the question, how has technology changed how we communicate with others? Next, using Chapter Three's method of philosophical reflection, I identify dimensions of our metaproblems with communicating amid technology. Chapters Three, Four, and Five offer openings for understanding and responding to our metaproblems. Chapter Six weaves together this study of communicators who find themselves in an ecos significantly changed by technologies, who (mis)understand their own being, being as bodies, and being *with*. This project does not end with complete answers to our questions or perfect solutions to our problems. The problems with human communication described here have no answer in the back of the book. Projects like this, adapting Gabriel Marcel's philosophical method, look around for a place to start trying to solve our problems. Amid our failures in communicating amid technology, where might we find ground to start again?

CONCLUSION

Even without technology as barrier, we struggle with problems in our interpersonal communication and in our efforts to dialogue. Technology may assist us in addressing some of these problems and allowing us to communicate better. Technology may also magnify our communication issues in familiar situations of feeling "absent present" as we simultaneously talk to this person and text another person or of feeling "alone together" in a room full of people scrolling through their feeds.[72] Our problems of communication are intrapersonal and interpersonal problems: we feel unseen, unheard, misunderstood, and sometimes alone. Marcel calls these feelings of isolation, uneasiness, or dissatisfaction signs of living in a broken world. People feel broken because they have lost "a sense of being," and they "refus[e] to reflect and refus[e] to imagine" situations distinct from what they currently are.[73] Often, communicators refuse to turn back and reflect on who they are in their world. This breakdown in intrapersonal communication spills over into divisions in interpersonal communication. In our broken world, our problems of communication portend impossibility of dialogue. If we cannot even transmit messages that are understood, we cannot ascend toward more-than communication: into a realm of communion, a realm of dialogue.

Problems in human communication cannot be fixed. A broken world does not invite scientific proposals that solve the problem. When human beings are broken, we need healing. Marcel suggest a way for healing: awakening to being. Healing our broken communication might begin in awakening to being. Awakening involves reflection upon (an action associated with seeing or looking upon) and coming into the presence of (a sensing the other through an act of receiving). Awakening is not a certain, clear strategy based on sharp enlightenment. Marcel called his autobiography *En Chemin vers Quel Éveil* (*On the Way to What Awakening*). This project adds to an ongoing discussion about Marcel's philosophy, a reflection on what awakening to being might mean for communication today.

Marcel did not offer authoritative explanations about what being in the world means,[74] he shared mid-journey reflections, "a quest," a *recherché*, "a search for, or an investigation into."[75] Philosophy, for Marcel, is an "aid to discovery rather than a matter of strict demonstration."[76] Marcel describes his own philosophical and life journey as an ongoing "awakening," an ongoing "searching."[77] This work, in the spirit of Marcel's work, does not set out to prove *that* awakening to being works, but rather to reveal *how* awakening to being might work.

In many of our relationships today—in workplaces, in families, in friendships—we stand with each other, all (or almost all) of us holding our cellphones in one hand. We are always connected through our technology, and yet, so many times, we feel as though the other person is unreachable, just outside of range of communication. We stand on opposite sides of a chasm, unsure of where the way across lies. This project sets out in search for the beginnings of a way across.

NOTES

1. The line from the poem is "Things fall apart, the centre cannot hold." William Butler Yeats, "The Second Coming," *The Collected Poems of William Butler Yeats*. https://www.poetryfoundation.org/poems/43290/the-second-coming.

2. After the killing of Breonna Taylor on March 13, 2020, and the killing of Ahmaud Arbery on February 23, 2020, and years of other deaths of black persons in the United States.

3. I ask myself if my topic of dialogue amid technology deserves addressing when much more pressing issues like systemic racism and individual/communal racist attitudes, actions, and philosophies in the U.S. call for immediate attention. Persons—much more qualified than I—have, are, and will be writing on this topic. Dwayne A. Tunstall, *Doing Philosophy Personally: Thinking about Metaphysics, Theism, and Antiblack Racism* (New York: Fordham Univ Press, 2013). Tunstall addresses how this particular philosopher, Gabriel Marcel, in his sociopolitical reflections ignores this very important way depersonalization as antiblack racism happens in this Western historical moment. Tunstall finds Lewis Gordon's study of existence addresses antiblack racism in Western experience. Tunstall identifies the particular instances in which Marcel does not acknowledge experiences of antiblack racism in his own country, France, and in American experiences. Tunstall identifies the limitations of using Marcel's reflective method in battling depersonalization happening as antiblack racism. One of the spaces we meet persons who think other than we do are our social media platforms. While this project does not address the racial biases, histories, privileges constituting each of our situation as communicators who use social media, this project contributes to a conversation about shared or unshared human experiences of being as bodies and being with others that may be talked about *after* and *while* acknowledging racism in the very structures, socialization, and philosophies about being human.

4. Martin Buber, *Between Man and Man*, translated by Ronald Gregor Smith (Mansfield Center, MA: Martino Publishing, 2014), 22.

5. Rob Anderson, Kenneth N. Cissna, and Ronald C. Arnett, eds., *The Reach of Dialogue: Confirmation, Voice, and Community* (Cresskill, NJ: Hampton Press, 1994).

6. Maurice S. Friedman, *Martin Buber: The Life of Dialogue* (London: Forgotten Books, 2015), 87.

7. Kwan Min Lee, "Presence, Explicated," *Communication Theory*, 14, no. 1 (2004): 27–50. Scholars have used the word "actual" to describe in-real-life, shared physical space and time in distinction from "virtual" for technologically mediated encounters. Ulrike Schultze, "Embodiment and Presence in Virtual Worlds: A Review," *Journal of Information Technology*, volume 25, no. 4 (2010), note 4, p. 446. Schultze notes that the word "real" also applies to virtual encounters, so he prefers the adjective "actual" to describe encounters between embodied persons who share physical space and time.

8. Nancy C. Baym, *Personal Connections in the Digital Age*, 2nd ed. (Maiden, MA: Polity Press, 2015), 1.

9. Joseph B. Walther, "Computer-Mediated Communication: Impersonal, Interpersonal, and Hyperpersonal Interaction," *Communication Research*, vol. 23, no. 1 (1996), 3–43.

10. Bernie Hogan and Anabel Quan-Haase, "Persistence and Change in Social Media," *Bulletin of Science, Technology & Society*, Vol. 30, no. 5 (2010): 309–315.

11. Gabriel Marcel almost exclusively uses masculine pronoun and the word "man" for person (Even his title of one of his books, *Man Against Mass Society* uses the word

"man"). I attempt to use more inclusive language with the words person, one, or woman. Citations from Marcel's texts will use the masculine pronouns, while explanations will move to more inclusivity.

12. Gabriel Marcel, *The Mystery of Being: Reflection and Mystery,* vol. I, translated by G.S. Fraser (South Bend, IN: St. Augustine's Press, 2001), 205.

13. Ibid.

14. Ibid.

15. Gabriel Marcel, *Metaphysical Journal,* trans. Bernard Wall (Chicago, IL: Henry Regnery Company, 1952).

16. Along with this diary, Marcel's primary philosophical works were "On the Ontological Mystery," in *The Philosophy of Existentialism,* translated by Manya Harari, 9–46 (New York: Citadel Press Kensington Publishing Corp, 2002); *Being and Having: An Existentialist Diary,* translated by Katherine Farrer (New York: Harper and Row, 1965); *Homo Viator: Introduction to a Metaphysic of Hope,* translated by Emma Crawford (New York: Harper & Row, 1965); *The Mystery of Being: Faith and Reality,* Volume II, translated by René Hague (Chicago: Henry Regnery, 1952); and *Creative Fidelity,* translated by Robert Rosthal (New York: Farrar, Strauss, and Company, 1976).

17. Meeting Gabriel Marcel, or any writer, does not happen in a vacuum. I encountered Marcel amid my doctoral studies in communication, but communication scholars made little to no reference to Marcel. I reference those few references at the end of this chapter. I turned to philosophers who had long been thinking about Marcel's works in meaningful ways. Thomas Anderson, *A Commentary on Gabriel Marcel's the Mystery of Being (*Milwaukee Wisconsin: Marquette University Press, 2006). Thomas Anderson's *Commentary on Gabriel Marcel's The Mystery of Being* offered many rich details about Marcel's significant Gifford Lectures and connected these lectures to Marcel's other works. Brendan Sweetman, "Introduction," in *A Gabriel Marcel Reader,* edited by Brendan Sweetman (South Bend, Indiana: St Augustine Press, 2011): 1–8. Sweetman's *A Gabriel Marcel Reader* highlighted themes throughout Marcel's works that gave me moorings for my own movement through Marcel's philosophical works. Richard Zaner, *The Problem of Embodiment: Some Contributions to a Phenomenology of Body (*Belgium: Martinus Nijoff, 1964). Jim McCown, *Availability: Gabriel Marcel and the phenomenology of Human Openness* (Missoula, MT: Scholars Press, 1978). Brian Treanor, *Aspects of Alterity: Levinas, Marcel and the Contemporary Debate* (NY: Fordham University Press, 2006). Dwayne A. Tunstall, *Doing Philosophy Personally: Thinking about Metaphysics, Theism, and Antiblack Racism* (New York: Fordham University Press, 2013); Donald F. Traub, *Toward a Fraternal Society: A Study of Gabriel Marcel's Approach to Being, Technology, and Intersubjectivity* (New York: Peter Lang, 1988); Jill Graper Hernandez, *Gabriel Marcel Ethics of Hope: Evil, God and Virtue* (London: Bloomsbury Academic, 2011). Philosophers Richard Zaner, Jim McCown, Brian Treanor, Dwayne A. Tunstall, Donald F. Traub, and Jill Graper Hernandez modeled for me applications of Marcel's ideas to our lived ways of being with each other.

18. Jean Wahl, *Philosophies of Existence: Introduction to the Basic Thought of Kierkegaard, Heidegger, Jaspers, Marcel and Sartre,* translated by. F.M. Lory (London: Routledge, and Kegan Paul, 1969).

19. *Mystery of Being,* vol. I, 26. For Marcel, "the broken world" referred to feelings of uneasiness in one's own life and one's dissatisfaction in one's relationships with others.

20. *Mystery of Being,* vol. I, 125. Marcel uses the phrase "being in a situation." See *Metaphysical Journal,* 137, for a definition. Marcel defines one's situation as "the unity of the situation appears to those 'involved' in it as essentially being a datum given [to consciousness] but at the same time as something that permits of and even calls for their active intervention."

21. Gabriel Marcel, *Philosophical Fragments: 1909–1914,* translated by Lionel A. Blain (South Bend, IN: The University of Notre Dame Press, 1965).

22. Gabriel Marcel, *Awakenings; Metaphysical Journal.* Marcel mentions his tendency toward idealist thought when a student at the *Sorbonne,* Marcel explored American

Idealists William Ernest Hocking and Josiah Royce. Denis P. Moran, *Gabriel Marcel: Existentialist Philosopher, Dramatist, Educator*. Lanham, MD: University Press of America, 1992. Hocking and Royce attempted to use an idealist approach for reflection on interpersonal interactions. David W. Rodick, *Gabriel Marcel and American Philosophy: The Religious Dimension of Experience* (Lanham, MD: Lexington Books, 2017), 3–9. Rodick offers insightful description of Marcel's exploration of idealism as promoted by Royce and then Hocking.

23. Gabriel Marcel, *The Existential Background of Human Dignity* (Cambridge, MA: Harvard University Press, 1963), 22.

24. Straus, Edwin W., and Michael Machado, "Marcel's Notion of Incarnate Being," in *The Philosophy of Gabriel Marcel*, vol. 17, edited by P.A. Schilpp and L.E. Hahn (Chicago: Open Court, 1984), 124.

25. *Mystery of Being*, vol. I, 4, *Metaphysical Journal*, xiii.

26. *Mystery of Being*, vol. I, 41.

27. Marcel, "On the Ontological Mystery," 11.

28. Gabriel Marcel, "The Broken World," in *Gabriel Marcel's Perspectives on The Broken World*, translated and edited by Katherine Rose Hanley, 31–152 (Milwaukee: Marquette University Press, 1998). In 1933, Marcel published the play "The Broken World" along with his essay "Ontological Mystery," which offers more characteristics of this broken world. Gabriel Marcel, *Mystery of Being*, vol. I, 18. Marcel titles a chapter of *Mystery of Being* "A Broken World," marking this as a major theme in his work.

29. Brian Treanor, *Aspects of Alterity*, 56. Scholars have not interpreted this "broken world" to point to a once-whole world that societal change has torn asunder.

30. *Homo Viator*, 80; Marcel, *Existential Background*, 128–133. In *Existential Background*, Marcel explores what human dignity means for those who do not believe that each life bears a sacredness because of the person bearing a "mark of the *imago dei*" or "image of God." Marcel suggests that nonreligious terms like "neighbor" and "fraternity" speak of a human dignity or sacredness inherent in each human being that deserves honoring.

31. For example, Marcel's insights into sacred dimensions of human living pair well with interdisciplinary scholar, Gregory Bateson. Bateson's works suggest an ontology connecting one's experience of beauty and participation in one's world. Noel Charlton extracts from Bateson's works, Bateson's acknowledgment of the sacred in systems of communication and interactions that create shared sustainable connections. Noel G. Charlton, *Understanding Gregory Bateson: Mind, Beauty, and the Sacred Earth*. Albany, NY: State University of New York Press, 2008, 157–158.

32. Gabriel Marcel, *Awakenings*, trans. Peter S. Rogers (Wisconsin: Marquette University Press, 2002), 192. From the 1940s, Sartre contrasted his own "atheistic" study of existence with Marcel's Christian existentialism. Marcel would express that he identified his philosophy neither as Christian nor *as* existentialist. Etienne Gilson, *Existentialisme Chrétien* (Librarie Plon, 1947). Gilson noted that some philosophers of existence, seeing no divine presence in the world, are identified as atheists, while other philosophers, like Marcel, are identified as religious, finding divine presence and meaning in experience. Marcel regretted that he agreed to be included in Gilson's work. Otto Friedrich Bollnow, "Marcel's Concept of Availability," in *The Philosophy of Gabriel Marcel*, vol. 17, edited by P.A. Schilpp and L.E. Hahn (Chicago: Open Court, 1984), 179.

33. Gabriel Marcel, *Awakenings*, 122–23. Marcel, *Being and Having*,15. Before 1929, although he reflected on spiritual presences, Marcel did not identify with a religion nor express explicit belief in God. In 1929, at the age of 40 Marcel experienced "an inner call to believe" in the Divine, a moment he described as "a birth . . . everything is different, a world which was there, entirely present and at last I can touch it."

34. *Homo Viator*, 7. Marcel uses the phrase "*homo viator*" as wayfarer or traveler.

35. *Metaphysical Journal*, xiii. Marcel stated his preference that his philosophy be described as a "kind of neo-Socratism."

36. Marcel, "On the Ontological Mystery," 44.

37. Richard Kearney, *Debates in Continental philosophy: Conversations with Contemporary Thinkers* (New York: Fordham University Press, 2004), 259. Kearney addresses the place of faith in philosophy by suggesting that faith or unbelief form one's "presuppositions" that may or may not be suspended during one's philosophical reflections. Even if one sees religious belief as a way of understanding the world, one can write as a philosopher, performing a philosophical hermeneutic study.

38. Marcel, *Existential Background,* 134. Marcel describes that one who does not believe in a divine power, may have a "poignant experience of the mystery inherent in the human condition." Gabriel Marcel, *The Decline of Wisdom,* translated by Manya Harari (New York: Philosophical Library, Inc., 1955), 19. Marcel describes the sacredness to being human that if ignored brings tragic consequences.

39. Gabriel Marcel, *Man Against Mass Society,* translated by G. S. Fraser (Chicago: Henry Regnery Company, 2008), 59–60.

40. Ibid., 61.

41. Rob Anderson, Kenneth. N. Cissna, and Ronald C. Arnett, eds. *The Reach of Dialogue: Confirmation, Voice, and Community* (Cresskill, NJ: Hampton Press, 1994); Ronald C. Arnett and Pat Arneson, *Dialogic Civility in a Cynical Age: Community, Hope, and Interpersonal Relationships* (Albany: SUNY Press, 1999); Rob Anderson, Rob, Leslie A. Baxter, and Kenneth N. Cissna, eds. *Dialogue: Theorizing Difference in Communication Studies* (Thousand Oaks, CA: SAGE, 2004); Ronald C. Arnett, Celeste Grayson and Christina McDowell, "Dialogue as an 'Enlarged Communicative Mentality': Review, Assessment, and Ongoing Difference," *Communication Research Trends,* vol. 27, no. 3 (2008): 3–25. Arnett et al. offer reviews of significant dialogue scholarship. Tineke A. Abma et al., "Dialogue on Dialogue," *Evaluation,* vol. 7, no. 2 (2001): 164–180. Abma et al. also offer a summary of key discussions about dialogue from diverse perspectives.

42. Richard L. Johannesen, "The Emerging Concept of Communication as Dialogue," *Quarterly Journal of Speech,* vol. 57, no. 4 (1971): 373–382; Ronald C. Arnett, "Toward a Phenomenological Dialogue," *Western Journal of Speech Communication,* vol. 45 (1981): 201–212; Michael L. Kent, and Maureen Taylor, "Building Dialogic Relationships Through the World Wide Web," *Public Relations Review,* vol. 24 no. 3 (1998): 321–324.

43. Some communication scholars Leslie A. Baxter, H.L. Goodall Jr., Michael J. Hyde, W. Barnett Pearce, Kimberly A. Pearce, and John Stewart contribute chapters in this landmark work on dialogue. Rob Anderson et al., *Dialogue: Theorizing Difference in Communication Studies* (Thousand Oaks, CA: Sage, 2004).

44. Thomas Anderson, "The Body and Communities in Cyberspace: A Marcellian Analysis," *Ethics and Information Technology,* vol. 2, no. 3 (2000), 153–158. Jim McCown, *Availability;* Rodick, *Gabriel Marcel and American Philosophy;* John E. Smith, "The Individual, the Collective, and the Community." *The Philosophy of Gabriel Marcel,* vol. 17, edited by P.A. Schilpp and L.E. Hahn (Chicago: Open Court, 1984), 337–349. Donald F. Traub, *Toward a Fraternal Society.* Brian Treanor, *Aspects of Alterity.*

45. Richard Johannesen, "Communication as Dialogue."

46. These scholars explore elements of dialogue within Marcel's work. Dennis Cali, "The Ecology of Presence and Intersubjectivity in Marcel's Philosophy of Gabriel Marcel," *China Media Research,* vol. 11, no. 2 (2015), 109–117. Richard A. Engnell, "Toward an Ethic of Evocative language: Contemporary uses of holocaust-related terminology," *Southern Journal of Communication,* vol. 66, no. 4 (2001): 312–322. Within the communication literature, Richard Engnell and Dennis Cali identify ways Marcel's ideas open up our understanding of communication. Katherine Rose Hanley, "Gabriel Marcel and Postmodernism," *Bulletin de la Société Américaine de Philosophie de Langue Française,* vol. 7 (1995), 122–147. Rosa Slegers, "Le point sur les I: Concrete and Philosophical approaches to Commitment and Writing in Gabriel Marcel's Work," *Renascence,* vol. 55, no. 3. (2003): 259–269. In interdisciplinary communication studies, Katherine Rose Hanley and Rosa Slegers find Marcel engaging dialogue through his plays. Thomas W. Busch, *Circulating Being: From Embodiment to Incorporation Essays on Late Existentialism* (New York: Fordham University Press, 1999): 38–39. Busch offers a

brief exploration of the distinction between intersubjective relations of dialogue and relations based on objectivism.

47. Sarah Bakewell, *At the Existentialist Café: Freedom, Being, and Apricot Cocktails with Jean-Paul Sartre, Simone de Beauvoir, Albert Camus, Martin Heidegger, Maurice Merleau Ponty and Others* (New York: Other Press, 2016), 323–325.

48. Thomas Anderson, "Communities in Cyberspace."

49. Bernard Gendreau, "Gabriel Marcel's Personalist Ontological Approach to Technology," *The Personalist Forum*, vol. 15. no. 2. *Proceedings of the International Conference of Persons, Santa Fe, NM, 1999* (University of Illinois Press, 2003): 229–246. Gendreau highlights Marcel's call to reflect mystery is particularly appropriate for people caught up technology-use to the detriment of reflection.

50. Dennis Cali, "Ecology of Presence," 113.

51. Dennis Cali, "Ecology of Presence," 115.

52. Donald F. Traub, *Toward a Fraternal Society*, 220–223.

53. Ibid, 211.

54. Gabriel Marcel, *Man Against Mass Society*, 184. Traub quotes this particular text by Marcel.

55. Leslie A. Baxter and Barbara M. Montgomery, *Relating: Dialogue and Dialectics* (New York: Guilford Press, 1996).

56. Gregory Bateson, *Steps to an Ecology of Mind* (Chicago, IL: University of Chicago Press, 2000).

57. Paul Watzlawick, Janet Beavin Bavelas, and Don D. Jackson. *Pragmatics of Human Communication: A Study of Interactional Patterns, Pathologies, and Paradoxes* (New York: W.W. Norton and Company, 1962).

58. Ramsey Eric Ramsey, *The Long Path to Nearness: A Contribution to a Corporeal Philosophy of Communication and the Groundwork for an Ethics of Relief* (New York: Humanity Press, 1998).

59. Brendan Sweetman, "Introduction." Brian Treanor, *Aspects of Alterity*. Jill Graper Hernandez, *Gabriel Marcel's Ethics of Hope*. David W. Rodick, *Gabriel Marcel and American Philosophy*, 17. Rodick summarizes this "resurgence of interest in the philosophy of Gabriel Marcel" particularly within philosophy.

60. Sarah Bakewell, *At the Existentialist Café*.

61. Julian Marías, "Love in Marcel and Ortega," in *The Philosophy of Gabriel Marcel*, vol. 17. edited by P.A. Schilpp and L.E. Hahn (Chicago: Open Court, 1984), 555.

62. Pat Arneson, ed., *Perspectives on Philosophy of Communication* (West Lafayette, IN: Purdue University Press, 2008). Jason Hannon, Ed., *Philosophical Profiles in the Theory of Communication* (New York: Peter Lang, 2012). Claude Mangion, *Philosophical Approaches to Communication* (Chicago: University of Chicago Press, 2011).

63. Ibid.

64. Susanne K. Langer, *Philosophy in a New Key* (Cambridge, MA: Harvard University Press, 1957), 3–4. Langer offered this insight into philosophy woven through the study of media ecology, culture, and philosophy.

65. Lenore Langsdorf, "Epistemology, Tropology, Hermeneutics, and the 'Essence of Language,'" in *Experiences Between Philosophy and Communication*, 169–190, edited by Ramsey Eric Ramsey and David James Miller (Albany, NY: SUNY Press, 2003), 177. Langsdorf describes hermeneutics as an "excavation, in contrast to deconstruction, [that] thematizes implicit facets of a phenomenon by articulating what else is present, along with its more evident constituents."

66. Gadamer, Hans-Georg, *Truth and Method*, translated by Joel Weinsheimer and Donald G. Marshall (New York: Continuum, 2004).

67. Marcel uses the term "technics" for technology. Chapter Two identifies what Marcel means by this term.

68. Marcel, *Homo Viator*, 138.

69. Richard Lanigan, *Speaking and Semiology: Maurice Merleau-Ponty's Phenomenological Theory of Existential Communication* (Tubingen: De Gruyter Mouton, 2013). Frank Macke, *The Experience of Human Communication: Body, Flesh, and Relationship* (Madison:

Fairleigh Dickinson Press, 2016). Algis Mickunas, "Maurice Merleau-Ponty," in Pat Arneson (ed.) *Perspectives on Philosophy of Communication* (Lafayette, IN: Purdue University Press, 2007), 137–162. Lanigan, Macke, and Mickunas point to more extensive communication scholarship on embodiment.

70. *Mystery of Being,* vol. I, 163. *Disponibilité* has been translated into English as availability or disposability and *indisponibilité* as unavailability or non-disposability. Marcel, bilingual in French and English, suggested that the English term availability did not fully describe what he meant by the philosophical phrase *disponibilité,* In his lectures in English, he would frequently use the French word *disponibilité* and describe what he meant by the phrase.

71. Thomas Anderson, "The Body and Communities in Cyberspace: A Marcellian Analysis," *Ethics and Information Technology,* vol. 2, no. 3 (2000). Dennis Cali, "The Ecology of presence and intersubjectivity in Marcel's philosophy of Gabriel Marcel," *China Media Research,* vol. 11, no. 2 (2015): 109–117. Bernard Gendreau, "Gabriel Marcel's Personalist Ontological Approach to Technology," *The Personalist Forum,* vol. 15. no. 2. *Proceedings of the International Conference of Persons, Santa Fe, NM, 1999* (University of Illinois Press, 2003): 229–246. Donald F. Traub, *Toward a Fraternal Society: A Study of Gabriel Marcel's Approach to Being, Technology, and Intersubjectivity* (New York: Peter Lang, 1988).

72. Sherry Turkle, *Alone Together: Why We Expect More from Technology and Less from Each Other* (New York: Basic Books, 2011). Kenneth J. Gergen, "Cell phone technology and the challenges of absent presence," in *Mobile Democracy: Essays on Society, Self and Politics,* edited by V.K. Nyíri (Vienna: Passengen, 2003): 103-114. Gergen offers describes this experience is feeling "absent presence."

73. "On the Ontological Mystery," 9. *Mystery of Being,* vol. I, 36.

74. Gabriel Marcel, *Tragic Wisdom and Beyond,* translated by Stephen Jolin and Peter McCormick, edited by John Wild (Evanston: Northwestern University Press, 1973), 30. Marcel intended his philosophy to "show" not "prove" what he meant.

75. Marcel, *Mystery of Being,* vol. I, 1.

76. Ibid, 2.

77. Marcel, *Awakenings,* 79.

TWO

Technology's Changes and Our Diminishing Relationships

Gabriel Marcel's Approach to Technics

Living from 1889 until 1973 in France, Gabriel Marcel witnessed and experienced significant technological change to societal and interpersonal environments. Because Marcel studied the human person, he would also attend to the human person living in a world that was rapidly and drastically changing due to technology. Marcel, who by profession was first a playwright (then a philosopher), pays attention to the scene that is set for any character study. This chapter explores Marcel's ideas about how this scene, this world, is impacted by the presence of technologies. His reflections on technology frequently turn to describe broad societal experiences of increased industrialization and bureaucratization in modern France. During Marcel's lifetime, drastic technological changes emerged as chemical and large-scale warfare in World War I, concentration camps then employed before and during World War II, and the atomic bomb. Marcel's twentieth-century *mise en scène* permeated by technology often meant tragedy for and between human persons. Marcel viewed modern France as a world broken by technologies.

A world broken by technology affects the persons living within that world. Technology affects how human persons live in their world, think about their world, and address others living *with* them in this broken world. Marcel questions, do technics principally improve or degrade human living? Although Marcel does not extensively describe how technology affects ways people communicate, his philosophy contributes valuable insight about how technology changes human interactions. Marcel offers a point for reflection, "the concrete relationship" between human persons and their technical processes.[1] Marcel's general philosophical approach is marked by attending to particular, actual experiences.[2] While offering sometimes generalized claims[3] about societal experiences of technology, Marcel also examines how technology has created change for the human person in her environment, thoughts, and relationships. Marcel identifies a diminishing of our feeling at home in our environments, of our value of life itself, and of the value of the other person's life. Our disconnections in and devaluations of life itself create a backdrop of discord for our being in the world.

Because Marcel offered a number of reflections on technology throughout his corpus, this project summarizes and weaves together Marcel's texts on technology. This chapter explores these questions: What does a philosophy of technology explore? What is Marcel's approach to technology and technology's impact on human interactions? I use Marcel's term "technics" to describe several topics Marcel addresses in his works. He does not call his own discussion of technics a philosophy. His reflection upon the human person in relation to her environment, herself, and other persons *sounds* as a philosophy. Marcel's insights on technics' ways of changing our human interactions provide important context for how persons communicate in these changed contexts.

This chapter begins by situating Marcel's philosophical approaches to technology within other philosophies of technology. Second, Marcel's general approach to technology is summarized through his definitions and descriptions of technics. Third, I note Marcel's brief comments on how technics may promote human living. Fourth, I extract Marcel's ways that technical developments bring people disconnections from their environments. Fifth, I identify Marcel's proposal that a technical view brings a diminished sense of life. Finally, I trace Marcel's insights into technical attitudes leading to devaluation of the other person. Marcel's description and evaluation of how people are changed by technology uncover an underlying feature to the relationship between the technology-user and her technology: "It's complicated."

PHILOSOPHIES OF TECHNOLOGY

In his study of how people and society understand, create, and use technology, Marcel joins past and contemporary philosophers. The word technology finds its definitional roots in classical Greek philosophers Plato and Aristotle differentiating craft knowledge from scientific knowledge.[4] Craft knowledge, *technē*, is understanding *how* to produce or create something.[5] Scientific knowledge, *epistēmē*, is seeking truth about the world through induction or deduction. *Technē* involves the "exercise of the craft," the producing of something "whose origin is in the producer."[6] Plato and Aristotle write about technical knowledge from a classical worldview of the world as cosmos, as *physis*, as the natural world. Within this worldview, to know the world means to be in harmony with that world.[7] Plato evaluates *epistēmē* as a higher form of knowledge than *technē* because *epistēmē*, rational knowledge, leads one to full understanding of the truth of the world and abstract reality. According to Plato, *technē*, is a lesser form of knowing.[8]

These classical Greek descriptions of technological knowledge contrast with modern definitions for scientific knowledge that depicts the one who has scientific knowledge of the world as having power over that world.[9] Modern sociologist Auguste Comte proposed that knowledge *progresses* through stages—or knowledge states—of explaining the universe. First, in a theological state, one believes in "arbitrary intervention" by "supernatural agents" as explanations for the universe.[10] Second, in the metaphysical state, one believes that "supernatural agents" are actually abstract forces "inherent in the different beings of the world." Third, in the positive (or scientific) state, one leaves behind studying origins or final causes of world phenomena to identify "the actual laws of phenomena" found in the world.[11] Within Comte's description of knowing, a modern evaluation of our ways of knowing emerges. One type of knowing, scientific knowing, emerges as a *better* way of knowing that has progressed *past* theological or metaphysical knowing.

Marcel's description of how the human person engages technology aligns less with modern descriptions of knowledge and more with the classical Greek descriptions of technological knowledge. According to Marcel, one could interpret the world from an objective, scientific perspective or from a metaphysical perspective. "Objective thinking" involves seeking clear definitions and precise analysis of phenomena in the physical world.[12] Marcel associated objective thinking to the scientific method in which one takes a spectator's perspective toward the object of study.[13] In objective thinking, one stands back from the phenomenon to describe the reality. An objective perspective seeks to discover precise descriptions of the material, physical world. Marcel described objective thinking as a valid but limited method for studying the world. Through objective study, one cannot know or describe "the spiritual content of the world."[14] A person who engages objective thinking cannot access the presence of being, transcendence,[15] or mystery.[16] A strictly objective, scientific knowledge of the world does not allow for experiencing or knowing phenomena in a metaphysical realm. Yet, the world bears dimensions beyond material, physical realms. Objective thinking does not take into consideration the *being*, or person, who reflects upon the situation.[17] Marcel promoted a reflective approach to the world that remains open to experiencing metaphysical phenomena like being, mystery, or the human person. With a reflective approach, one can study phenomena in metaphysical dimensions the world, "from the standpoint of metaphysical knowledge the world remains the site of uncertainty, the reign of the possible."[18] The world, as a site of possibility, is a world open to the presence of being and transcendence. Marcel offered his reflective approach as *an* interpretation of how one might think about technology and the world.

Along with definitions for technological knowing, scholars have offered philosophical evaluations on how human persons create and engage technologies to improve or diminish the human situation. From ancient Greece, philosophers were weighing in on the promise or problems offered by technology. Plato's Myth of Theuth evaluates writing as a technology: with one speaker promoting writing as an aide to our knowledge of the world and the counter speaker suggesting memory will suffer because of this new technology of writing.[19] Modern philosophers—precursors to Gabriel Marcel—like Comte or Karl Marx,[20] optimistically suggested that technological advances mostly brought human progress. Scholars from various fields including but not restricted to philosophy, sociology, psychology, and physical sciences offer insights into *how* human persons engage or interact with technology.[21] Joining his contemporary twentieth-century philosophers,[22] Marcel offered his evaluation of how technology has changed the human condition and experience.

Some discussions of technology's intersections with communication emerge within the interdisciplinary field of study media ecology. A media ecology perspective on technology attends to human interactions through, amid, or situated in technologies as impacting or impacted by human communication and culture. Neil Postman described media ecology as a study of "media as environments."[23] Media ecology examines "how media of communication affect human perception, understanding, feeling, and value."[24] A media ecology approach to technology weaves together philosophy, communication, sociology, psychology, history, and cultural studies as reflections on how media or technologies affect the human condition. The "medium is a technology within which a culture grows."[25] Television, radio, and social media are examples of technologies within which a culture grows. Lance Strate in his exploration of the meaning of media ecology notes that the study of media ecology is study of *persons who* study the human world with a similar approach.[26] Instead of asking, "*what is media ecology*," one should ask "*who is media ecology?*"[27] Media ecology scholars offer reflections on philosophies of technology offered by various twentieth-century thinkers—i.e. Marshall McLuhan, Walter Ong, Neil Postman, Jacques Ellul, Lewis Mumford, Martin Heidegger, Elizabeth Eisenstein, and others.[28] Media ecology involves exploring media as environments as a lens for understanding human experience.

Communication scholar Dennis Cali identifies Gabriel Marcel as contributing descriptions of technology that fit within a media ecology perspective. Marcel did not identify his reflections on technology within a field or school. Strate, in offering his insights into media ecology at a particular time, makes explicit the opportunities for further answers to "Who is media ecology?"[29] Strate has characterized media ecology as "a field," as ground for ongoing excavations of meaning.[30] As such, media ecology offers multiple openings for shared approaches to understanding how the human person uses, relates to, and lives amid technology. At the end of this chapter, this work continues initial explorations of overlaps between these discussions within philosophies of technology and media ecology and Gabriel Marcel's study of human experience of technology.

Several philosophy scholars, Thomas Anderson, Bernard Gendreau, Dwayne A. Tunstall, and Donald F. Traub along with communication scholar Dennis Cali offer significant insights connecting Marcel's philosophy to experiences of technology.[31] Marcel's interpretations of technology's effect on human *communication* have not been extensively explored.[32] This project extends these discussions by offering more details on Marcel's approach to technics and exploring ongoing implications of this approach for our interpersonal communication.

Marcel's insights into technology were published before and alongside other significant reflections on technology offered by twentieth-century scholars.[33] Marcel's considerations on technology appeared first in 1933 in his essay "On the Ontological Mystery" and then in 1935 in his book *Being and Having*. His 1940 *Creative Fidelity* and 1945 *Homo Viator* included some reflections on the effects of technology on a person's self-understanding or relations to others. *Man Against Mass Society*, published in 1951, constituted Marcel's primary text on personal and communal relations to technology. His book *The Decline of Wisdom,* published in 1953, considered how technical progress has affected human persons' ability to live wisely. Marcel's 1953 work included chapters named "The Limitations of Industrial Civilization" and "Remedies for Dehumanization of the World." In 1963 Marcel delivered a lecture in Heidelberg named "The Sacral in the Era of Technology."[34] As well, in some sections of his 1963 Harvard University William James lecture,[35] Marcel addressed how the human person had been affected by technical progress. Marcel's interview with philosopher Paul Ricœur, recorded in *Tragic Wisdom and Beyond* in 1971, offered Marcel's final reflections on technology. As a Parisian man from a position of privilege in an aristocratic family, Marcel weighed in on general ways that technology had affected how people experience small scale and broad societal scale technologies in Western European environments.

MARCEL'S APPROACH TO TECHNICS

Marcel's words used to describe technology are complicated by the differing translations and changing interpretations throughout the years he was writing. Thomas Anderson claimed that Marcel most frequently used the French word "la technique" to describe interactions with technologies.[36] Some English translations converted the word "la technique" to technology.[37] Other English translations of Marcel's French writing used the words technique or technical.[38] In his lectures that he delivered in English, Marcel used a mix of the words technique, technical, and technocracy. Translators Jolin and McCormick indicated some overlap between Marcel's use of the words technology and technique.[39] Marcel sometimes used the word technology to specifically refer to materials used as instruments.[40] Anderson suggested that the English word "technic" is the most accurate translation of Marcel's word "la technique."[41] According to Anderson, the word technics encompasses the complex phenomena of technique and technology along with the descriptor technical.

Marcel did not always provide clear, consistent definitions for his use of the terms: technics, technique, or technical. Sometimes he explicated his phrases through descriptions or examples. Marcel uses the term technics for a range of human actions and material products. Technic also refers to the product created through scientific or technical knowledge, similar to Aristotle's *technē* of bringing something into being.[42] Marcel uses technics for "things constructed by human beings as instruments to attain desired goals."[43] Technics are things, machines or systems using scientific knowledge, created by human persons for a particular purpose. Technics are the product of reasoning and organizing functions. Technology is "human reason insofar as it strives to manage, so to speak, the earth and everything living within it."[44] Technology also refers to specialized, technical knowledge that can be taught to others.[45] In these descriptions, technics implies procedures, reasoning, and taught skills and also refers to material things or objects that are methodized or structured by human reason. Marcel sometimes used the word "technique" to describe how people *use* technology. He specified his definition for the word technique, as a "a group of procedures, methodically elaborated, and consequently capable of being taught and reproduced, and when these procedures are put into operation, they assure the achievement of some definite concrete purpose."[46] For example, people use "techniques of degradation" when using persons as a means to serve a national or economic purpose.[47] The concept technics deals with human reason acting upon and through material things or systems and human skills creating and organizing. As well, technics include processes, systems, machines, materials, functions and attitudes shaped by and shaping persons living in the world.

Marcel also used the words technocratic and technocracy[48] to describe attitudes, processes, and systems in technological twentieth-century Europe and Americas. One with a technocratic perspective views the natural world as comprised only by physical matter and forces.[49] In a society dominated by technics, the human person is defined by his function or usefulness in a technical system.[50] For Marcel, "technocracy" is a system of total commitment to techniques, governed by most-efficient technical processes, at the service of a nation, company, or economic endeavor.[51] Marcel identified the Nazi government in Germany as one such technocracy because they engaged "the massive and systematic employment of such techniques [of degradation], particularly in concentration camps."[52] In a technocracy, technics are prioritized, and the human person does not matter. Thus, the underlying spirit of a "technocracy is inevitably nihilist."[53] Marcel wrote on technocracy in 1950s, in the presence of the aftermath of World War II which witnessed countries engaging technocracy for national territorial pursuits. Marcel condemned technocracy promoted both by the U.S. and by the Soviet systems because the techniques no longer served the person but the system itself.[54]

Marcel's approach to technology sounds amid broader studies of and definitions for technology from a media ecology approach. A media ecology perspective describes technologies as both shaping the media and being shaped by environments. A "medium is a technology within which a culture grows"[55] and media as environments "consist of techniques as well as technologies, symbols as well as tools, information systems as well as machines."[56] Technology includes techniques, material technologies, machines, and information systems.

Marcel's contemporary scholars, Lewis Mumford and Jacques Ellul, offered definitions for technology that provide interesting context for and similar approaches to Marcel's definitions. In his descriptions of how ages in human history have been marked by varying technological ecologies, Mumford used the word "*technics*" as a broader description for human interaction with technologies. Technics are the "translation into appropriate, practical forms of the theoretic truths."[57] This translation of scientific ideas into created tools happened through "invention and experimental adaptation."[58] One must learn the "laws" that govern behavior of tools, materials, and persons as these interact within environments.[59] Because technics are "constant instrument of discipline and education,"[60] technics function as ordering those who use technics, the materials, the environments, and the interactions between all participants. Mumford's term technics parallels some of Marcel's descriptors for technics as "things constructed by human beings as instruments to attain desired goals."[61] Another scholar, Jacques Ellul, used the word "*technique*," in distinction from French word "*technologie*," translated into English as "technology."[62] Ellul offers *technique* as "the technical phenomenon," the "efficient methods applicable in all areas (monetary, economic, athletic, etc.)"[63] or "the totality of methods rationally arrived at and having absolute efficiency (for a given stage of development) in every field of human activity."[64] For Ellul, *technique* as a "sociological phenomenon," is in relation to and underlying all relations within society.[65] Ellul suggests that *technique*, and not work, sets all value in society.[66] Ellul's description of *technique* sounds similar to Marcel's description of technology as human reason managing, ordering materials,[67] or of technique as the coordinating "procedures" enacted for purpose.[68] Although Marcel does not cite either Mumford nor Ellul, Marcel's words sound in sync with media ecology characterizations of technologies.

In scholarship about Marcel's writing, media ecology or current broader definitions for technology may influence the translations of Marcel's original ideas. Acknowledging complexities in the use of the word technology in Marcel's works, this project defers to Anderson's approach of calling Marcel's general name for technology, technics. When sharing Marcel's reflections about technics, this project includes the terms the particular translators used for Marcel's phrases.

Marcel's evaluations of technology emerge against a historical environment scarred and devastated by scientific and technological "progress" emerging as the Great War of 1914 and then as World War II. France in the Interwar period 1918–1940 included a complex interplay of national and personal responses to the trauma of the Great War of 1914. A modern assumption, accompanying belief in the authority of science, was that with increased scientific knowledge and discovery, humans and human society would progress. Modern science promised to control the world and thus improve lives.[69] Modern education also encouraged this belief in the power of the states to improve or strengthen cultures and make economies grow.[70] As modern individuals progressed in knowledge of the material world, society was to also to improve.

In the early twentieth century, while nation states, economies, and society seemingly progressed, persons also experienced increased distress in the name of this progress. During and after the Great War of 1914, French people questioned self-evident values like scientific success or protection of national interests as being actual values for human persons.[71] Some modern science used at the service of expanding modern nation states had resulted in the destruction of an unprecedented number of lives in Western Europe, the opposite of progress. The Great War, serving the growing nation state, brought tragic consequences to the lives of modern men and women. Henri Bergson commented at the start of the Great War that armies had "made their barbarism 'scientific,'" and that war revealed the power and effects of technological weapons to destroy human life.[72] Marcel's reflections on technology emerge against this background scarred by war. Thus, Marcel's insights into interactions between technics and human persons highlight predominantly negative effects on human persons. Before addressing how technics have introduced adverse changes for human persons, I summarize Marcel's identification of how technics bring advantageous—albeit limited—changes for human persons.

TECHNICS *MAY* PROMOTE BETTER HUMAN LIVING

Marcel acknowledged that certain uses of technics promote openings for innovative human actions. Technics may be used at the service of human persons. The development of technics also provides opportunities for improvements of particular human situations. Technics reveal opportunities for human creativity. The creation of technics can be an ennobling process. In artistic fabrication of technics, participants engage inventiveness through the reflective act of creating.[73] Marcel suggested, "Where there is creation there can be no degradation, and to the extent that technics are creative, or imply creativity, they are not degrading."[74] Technics manifest the creativity of its maker. Marcel proposed that creative generation is not itself a degrading action. Being inventive, the technic producer and the scientific researcher can be oriented beyond himself. The development of technics offers an opening for reflection, invention, and production. According to Marcel, although the technic inventor may infuse his actions at the service of creating a good, the technic-user more commonly unreflectively consumes technics. The intention of the person engaging technics imbues technics with its value at the service or in degradation of the human person. For Marcel, the reason for creating technics shapes how the technic benefits or harms the human person.

Marcel recognized a few ways technics brought positive changes for human life. Although Marcel thought the term "progress" was not an apt descriptor for all technological advances, he conceded that progress could be applied to situations in which technics provide new medical methods or improved housing and food opportunities for disadvantaged persons.[75] If technics are used to better serve human persons or communities, then he granted, "technical progress in the strict sense is a good thing, both good in itself and good because it is the incarnation of a genuine power that lies in human reason."[76] Technics manifest human reason applied to improving the human condition particularly in better transportation, communication, and connections between communities.[77] Better connections allow for and promise ongoing sharing of intellectual and technical goods via communication and transport channels. If technics are used to improve the lives and well-being of persons, then, Marcel granted, technics used in service of the human person could be considered good.

Although Marcel viewed promise in technical progress, he typically followed descriptions of technics' promise with qualifications about technics' effects on the human person. For example, he acknowledged that although "[T]echnology is good in itself . . . technology [could] be put to the wrong use."[78] The human person using the material technology may use technology in ways that harm human persons. In 1951 Marcel offered this insight into the complex relationships between technics and human persons,

> The realm of the technical, as thus defined, is not to be considered as evil in itself; if we think of it in itself, as I have already said, a technique is rather something good or the expression of something good, since it amounts to nothing more than a specific instance of our general application of our gift of reason to reality. To condemn technical progress is, therefore, to utter words empty of meaning. But from the point of view of truth, what we must do is not to cling to our abstract definitions but rather ask ourselves about the concrete relationship that tends to grow up between technical processes on the one hand and human beings on the other; and here things become more complicated.[79]

One should not reductively interpret the technical realm as "evil," harmful for human persons. Marcel recognized intrinsic good in technics as expressions of good inventions. Technics manifest an "application of our gift of reason to reality." Marcel suggested avoiding generalizations about technics: either concluding that technics are *always evil* or that technics *guarantee* progress. Thus, he invited reflecting on the "concrete relationship" that typically emerges between technical processes and human beings.[80] This relationship between human beings and their technics is not a simple relationship. Complicated relationships invite careful reflection. Marcel attends to human experiences in his contemporary and recent history in his Western European setting—where technics and human persons cohabit.

Technical inventions and increased technologizing have brought and continue to generate changes for human persons in their world. While technics provide opportunities for human creativity and improvements, Marcel also saw technics brought about changes with negative effects on human persons. Marcel studied changed relationships between the human person and his environment, technics, and other persons. The majority of his reflections on technics address ways human interactions with technics contribute to modern experience of the broken world. For Marcel, the broken world, marked by technical change, includes a general degradation of human persons. The next three sections address the ways technics have brought negative changes in human relations with environment, with technics and oneself, and with other persons.

TECHNICAL DEVELOPMENTS BRING DISCONNECTION FROM ENVIRONMENTS

Technics have brought significant change to human relations within their environments. Marcel's philosophy of technics considers how technological progress has created degrading changes to human relations with their environments. Community members have experienced transformations to ways of being in the world. Environment indicates both physical and social milieus. Marcel differentiated two types of environments for human persons, "the natural environment" and "the technical environment."[81] In a natural environment, found in pre-industrialized communities, one is a "craftsman," active participant in their work, while a technical environment, marked by mechanization, renders work "artificial and inhuman."[82] In this technical environment, one's contact with nature happens through mechanized processes and techniques. A natural environment offers "a climate of presence and sympathy," in which the human person is in harmony with his world. Technics have altered the way the human person dwells in and interacts with his world. Technical developments threaten an ideal experience of being human, of living in an internal and external harmony with one's world.[83] Marcel suggested, "technical progress, considered not in itself, not of course from the principles which made it possible, but incorporated into the daily life of individuals, has not been effected without the loss of human substance."[84] In this loss of human substance, persons experience discontinuity, dislocation, and increasing insecurity about one's own life.

Increasingly technical environments disrupt persons' daily rhythms. Marcel noted a "close connection between acceleration of the rhythm of life and the appearance of a humanity which is inwardly more and more impoverished."[85] For example, following timetables for working at factories, persons feel pressured to always increase their speed in production. In technical environments, persons follow a mechanized schedule for production. As one lives a more accelerated life, one experiences a lessening of one's internal well-being.

Marcel noted that, before the technical methods of the industrial age, time was felt differently. He gave the example of pilgrims from the Middle Ages who physically walked across lands en route to visit a place of significance. These pilgrims experienced time differently from modern workers. Pilgrims felt the passing of time as a "slowness of progress [that] . . . was linked to a feeling of veneration."[86] Now the demand to produce more and work faster changes the pace of life for the modern worker. Acceleration experienced in a technological age is accompanied by a loss of the "sense of the sacred."[87] Pilgrims in the 1300s felt the passing of time as a slow progression linked to veneration. Modern industrial workers, disconnected from life's natural rhythms, seek speed in production. In modern environments, time serves production and is no longer experienced as a natural rhythm of life. One of these losses is the loss of being able to live life slowly.

Technical progress in the industrial revolution involved experiences of dislocation as mass numbers of people moved from country settings to towns and cities. As modern communities experienced mass population movements due to war or totalitarian regimes, more people experienced that this "vital link [was] broken between man and his environment."[88] The modern worker, "technical man," displaced from his homeland, moves to a city, finds a job and has become the "one-who-can-make-shoes."[89] Building their self-worth from how they view themselves, human persons who see themselves as producer-of-work will value themselves on their level of production.[90] In a technical environment, the displaced person is reduced to what he can produce.

According to Marcel, as people moved into cities, workers left natural daily cycles of sunlight and dark and seasonal changes. One's separation from the natural cycles becomes a "dislocation . . . between man and life."[91] People who have moved into cities live a life no longer governed by seasons. Electricity creates unnatural cycles for persons, disrupting natural sleep or rest patterns. One can work through the night, through the heat or cold, like a machine whose one purpose is to produce. When one person working for the factory stops producing at top level, she can be replaced like a machine part with another worker. Life includes connections to nature: its time cycles, its spaces, and one's feeling harmony living in this world. In this dislocation, the city-dweller "ignore[d] more and more systematically his condition as a living being."[92] In a modern era, people can work through the night and live in wholly artificial environments disconnected from the natural circadian rhythms and seasons.

The modern person displaced in or by a technical environment feels insecurities about life itself. Modern science promised that this increased control over environments would bring fewer fears about being harmed by nature or situations.[93] For example, if people could build shelters from storms or protect themselves from animals, they would have increased power over their environments. As well, people continue to believe in technics to solve societal or personal problems. In the increasingly technological environments, Marcel did not find this increased sense of security, but rather witnessed growing numbers of people experiencing fear. Although improved technological methods seemed to increase one's control over natural situations, people's power over their own anxieties had not diminished.[94] Some of these fears and anxieties become "despair" due to "recognition of the ultimate inefficacy of all technics."[95] According to Marcel, while technics promise(d) increased control to its users, technics are accompanied by increasing fears of losing control of life itself.

Fears of losing control alter how one values and views one's own life. Marcel observed, "[A]s preoccupation with security begins to dominate human life, the scope of human life itself tends to be diminished. Life, as it were, tends to shrink back on itself, to wither."[96] Members of communities had lived through mass movements to cities, home displacements, and war destruction. People were increasingly worried about their own security in the world. Marcel connected this increased concern with a feeling that life itself also has limits. The fearful modern technical person's chief preoccupation becomes one's commitment to self-preservation.[97] The modern technical individual continually calculates any risks to his own life. For example, the modern individual calculates whether it is even worth bringing a child into this world.[98] Marcel called this calculating, preoccupied, reductive approach to life, a technical perspective.

Technical progress changes the ways human persons relate to their own environment. According to Marcel, increasingly technical settings brought dislocation from previously natural environments. Individuals working in modern technical settings feel disconnected from particular lands and places. Marcel witnessed growing feelings of insecurity about one's own life. One's positioning in the world has been altered through dislocation and discontinuity. In a technical world, one feels out-of-place from one's surrounding environment. Marcel related this feeling of displacement to a feeling of alienation from others in the world. This technical progress has also shaped how the modern person reflects upon the world: one's own internal world and the world around him. Within a-contextual settings of technical environments, one's own relation to technology has also shifted.

A TECHNICAL VIEW BRINGS A DIMINISHED SENSE OF LIFE

A technical perspective on life also shapes how one relates to one's own life and to the technology present in one's life. Marcel's reflections in 1933, 1935, and 1951 on the effects of technics on the person sound enigmatic for those historical contexts and prophetical for contemporary settings. Marcel, writing his 1935 work, does not address relations to technics as understood in the early twenty-first-century experience of current technologies. However, Marcel's reflections on how persons possess and relate to one's things surprisingly still apply to contemporary human experiences. According to Marcel, in technical settings, persons experience having an increased dependency on technics, a limited sense of the self, no belief in life-after-death, and no awareness of the presence of mystery.

Marcel acknowledged that although technics make one's life easier, they also render one more dependent upon technics. A person dependent upon technics centers her attention solely on external objects and interactions. As potential consumers, continually addressed through advertising, they become reduced to consumers.[99] In his 1963 reflections, Marcel already witnessed technology consumers tethered to new technical developments and thoughtlessly caught up in the emergence and use of those technics. Marcel indicated "The more a man becomes dependent on the gadgets whose smooth functioning assures him a tolerable life at the material level, the more estranged he becomes from an awareness of his inner reality."[100] This dependency on technics alters one's internal life of reflection and relations.

Most of our relationships with technics are relationships of consumption. Consumers relate to technologies as consumable things and as possessions. Marcel discussed distinctions between how one relates to objects through *having* or through *being*.[101] These distinct relations affect the one-who-relates to objects. In one's relationship with things, the things have a certain effect on oneself: "[O]ur possessions eat us up . . . when we are in a strange state of inertia in face of objects which are themselves inert."[102] The more a person is "attached to these things," the greater the power[103] that the things exercise over the attached person.[104] One may *have* or possess something, but when the thing begins to influence the person, one loses power over that which one possesses. In a world of technics, "he is increasingly incapable of controlling his technics, or rather of *controlling his own control*" and soon feels "at the mercy of his technics."[105] Marcel describes, for the people in 1930s, persons who have not been able to control technics or even controlling how they interact with technics. People dependent on their "gadgets" live in a "process of automization," thoughtlessly focused on their material possessions and unaware of their own inner world.[106] In this "automization," or loss of control, one could become bound to technics. Idolatry—involving automatic actions and unreflective consumption—signals and perpetuates selfishness, or what Marcel called an "autolatry: worship of self."[107] "Autolatry" means withdrawing within oneself in unreflective self-concern, a "degrading" or less-than human way of being in the world.[108] People in a technical environment seem incapable of identifying the role or power that technics assume over their own lives.

When one is fixated on an external technologized world, one may lose sight of one's inner world and undervalue one's own worth as human being. Marcel notes, work in a technological world leads people to compare themselves with their machines.[109] In these comparisons to his own products, a person "undervalues himself in comparison with the far more precise and effective apparatus which his technical skill has perfected."[110] In this comparison, a worker finds he is less effective than the machine, the product is more valuable than the one-who-produces it. The worker's health is secondary to the worker's production. One of Marcel's main hesitations about technics was that technics altered one's understanding of what it means to be a person who is both bodied and spiritual. Technics bring the person "a certain degradation of one's spiritual level."[111] Marcel clarifies that one need not be religious to understand his proposals for spirituality and being.[112] Marcel offers insights into being human. One's spiritual level could include one's search for inner peace, one's sense of mystery or auras, or that inner realm that goes beyond one's mind or one's body. Technocratic attitudes—privileging one's efficiency over one's bodily and spiritual dimensions—are a degradation of being human.[113] In this reductive comparison, one's spiritual dimensions are being ignored. Thus, Marcel warned against devaluing the bodily or spiritual aspects of the human person. In a technical world, one evaluates one's own worth against one's machine's productivity and one loses sight of one's own spiritual dimensions.

Marcel associated technical attitudes with a loss of appreciation for mystery or presence in life.[114] The "technical man has lost his awareness of himself and awareness of transcendent elements."[115] Persons remain fixated on physical and tangible dimensions of life. This lack of awareness of transcendence is connected to experiences of a broken world. People in a technological era attend strictly to material problems and dismiss the presence of mystery—even in their own inner worlds—as unreasonable. Commitments to strictly scientific thinking, even in the study of the human person, are connected to "devaluation of the person," a dismissal of presence of mystery and being in human life.[116] As Paul Ricœur noted, Marcel finds in human experiences in technological world, "a certain core of the sacred in man has been violated."[117] Marcel, consistently throughout his work, called for attending to the person as an integral whole: as someone open to—although not necessarily believing in—a life beyond the physical life.

Marcel argued that strictly technical attitudes affect how one views death and consequentially how one views life. Before the modern era, many people held a belief that the human spirit lived on after death, in a mysterious—beyond explanation—way.[118] Yet in the modern era, death of a loved one is presented as a "raw fact, like a dislocation of some piece of mechanism."[119] One's death is an abrupt end to one's presence in the world, a piece of a work or societal machine that has disappeared and become nothing. While some modern proponents suggest that the belief that a person lives on in some form after death is unscientific, this belief actually deals with experiences *beyond* scientific analysis.[120] In the era of technical progress, a modern approach to death is that death signals one's absolute end.[121] An objective approach to death proposes that because we cannot know with certainty what happens after death, life after death cannot objectively be verified nor confirmed as being real. Marcel highlights how this scientific conclusion about the possibilities for life after death precludes other interpretations of the mystery surrounding what life after death is or might be.

Marcel correlates views of life after death with views of life in this world. In the nineteenth century, people had suggested that scientific proposals for life would improve general well-being. According to Marcel, scholars predicted that as more people ceased believing in an afterlife, "life in this world would be more and more lovingly taken care of."[122] If an earthly life is all a person has for living, then the investment in that earthly life should greatly increase. However, Marcel observed, the opposite interpretation of life seemed to be prevalent in twentieth-century France, "life in this world has become more and more widely looked upon as a worthless phenomenon."[123] Marcel suggested that viewing life after death as impossible correlated with considering life itself to be replaceable or even disposable. According to Marcel, one's view of death reveals the value one assigns to living.

His view of the human person as open to belief in transcendence constituted a leitmotif throughout his philosophy of existence *and* his approach to technology. In the modern technological era, people placed their hopes in technical advances and solutions in their physical world and Marcel concluded that this misplaced hope had led to human tragedies. According to Marcel, persons had lost sight of the human person as being more than just a machine or as actually bearing spiritual dimensions. For Marcel, this technical view of the human person as being strictly material, producer, functionary-in-a-workplace bears real, and sometimes tragic, consequences for how one views other human persons.

TECHNICAL ATTITUDES LEAD TO DEVALUATION OF OTHERS

Technics and the technical view of life shape one's approach to *other persons* living in shared environments. People misinterpret and misunderstand who the other person is and what that person is worth. Marcel notes how technics have affected our view of this human life being sacred and gift

A technical view of the human person does not include an acknowledgment of the sacred in the human person. Marcel suggests, "in a world where technology enjoys absolute primacy, a desacralizing process inevitably sets in that is directed against life and all its manifestations."[124] Being bears a dimension that goes beyond these facets we grasp with our scientific knowledge. Before phenomena bearing transcendent dimensions, one stands before an "absolute, impassable gulf which opens between the soul and being."[125] All beings bear this transcendent dimension, before which one stands. Marcel links this description of bearing transcendent dimensions with being sacred.[126] A being that is sacred is irreducible to one's own mental categories, that which is sacred, must be stood before. Marcel describes persons who appear to one as sacred beings, "*mirabilia*" those who inspire wonder.[127] In wonder, one gazes upon the other, recognizing something worthy of wonder, and acknowledging it through one's verbal or nonverbal posture, eye contact, or words. One's evaluation of life as being anything less than sacred, brings consequences in how one evaluates the life of each human person.[128] To diminish any dimension of being a person is to desecrate, to degrade, to make less sacred that person. Acknowledgment of this life as being sacred fills the one who is being acknowledged and the one who acknowledges with attitudes of reverence for.

Marcel identified in his milieu a lack of belief that each human life is sacred. If one measures the value of a person's life according to that person's productivity in society, the other's life—with slower or no productivity—may loses its significance as being sacred.[129] Amid increased fears and anxieties, life itself seems insecure. For many persons in technical settings, life is no longer "a gift to be handed on, [but more] as kind of incomprehensible calamity like a flood."[130] If the value of one's own life wavers under the primacy of technics, a person feels uncertain about—or burdened by—bringing another life into this devalued existence.

Along with this reductive view of the person, in the world broken by technology, people privilege material production, wealth, and success over being. One of the results of privileging material wealth is increased self-centered actions. In a focus on distribution of wealth, people have a sense that the other's gain is one's own loss. Persons experience "an inner inertia" of feeling frozen by fear and anxiety about one's material resources being threatened.[131] Marcel noted how no "technique which will permit us to free ourselves" from this fear.[132] This fear leads people to fence themselves off from others to protect their wealth, possessions, and life.[133] Fenced off from the other, one has no "room" for the other within one's world.[134] These inner barriers to the other within one's interior world translate into real barriers in one's material world. If one cannot see the other person as being someone who also needs, who *is* there, in need next to a person, one is incapable of relating to the other as living, as irreducible, as a sacred person. Thus, a society that privileges material dimensions of the person to the exclusion of any spiritual dimension functions "radically against intersubjectivity."[135] Reductive approaches to the other lead to misperceptions of the other as being less than human or as being quite un*like* one another.

One mentally moves toward this view of the human person as less-than-sacred when one performs an abstraction of the other person. Marcel named this attitude "a spirit of abstraction" toward the other person.[136] One performs an abstraction by only focusing on certain aspects of the human person: their mind, their production, their contributions, their body, or their wealth. In a modern world, education and training prepare individuals for production or functions within a technological world.[137] This calculating view of the other person as abstract entity translates into work and societal organizations. A person or organizations with a technocratic perspective views the other person as a "mere unit of production and judge[d] his worth only in terms of productivity."[138] One sees the other for their output. When a person no longer produces, his value disappears and he could be eliminated.[139] When a person stops functioning or producing, the technocratic leaders replace persons like disposable parts of a machine. In this reduction, one becomes "the abstract individual all of whose 'particulars' can be contained on the few sheets of an official dossier," an abstract entity "ticketed, docketed, labeled" stripped of body.[140] In this search for better technological systems, the human person is listed as a part serving (or failing to serve) technological productivity. Technological environments that privilege mental achievements over all other human actions lead to "degradation" of the human person.[141]

Misunderstanding of the human person includes reducing the other person to abstractions, a name listed and identified by the state for work in a factory or military service.[142] If a person is simply an abstract concept, he is disposable. A disposable person can be sent to fight—and die—in wars, a nameless number. Marcel called these strategies of using human persons as if reduceable to function, "techniques of human degradation."[143] Techniques of degradation were used in the twentieth century (and most tragically continue into the twenty-first century) to create and use wars, forced labor camps, concentration camps, and atomic weapons.[144] These techniques are used in complete disregard for the human persons affected by the technics.

These problems of misperception in the broken world mean people ignore the sacred in each life, feel that the other's presence is a threat to one's own presence, and reduce the other to an abstraction that (no longer a who) does not deserve acknowledgment. Our misperception of others is founded on technical views of and attitudes toward the other person. These misjudgments lead to people ignoring the irreducible presence of the other person. A sacred dimension in every human being communicates a human dimension that cannot be reduced to productivity or usefulness. A strictly technical or objective approach to other people leads to self-centeredness and a breakdown of our ground for meeting the other person, for intersubjectivity. The person consumed with his own technics, frozen in his selfishness, no longer has room for other persons. Fearful for his own life, his own productivity or his own technics, he faces problems brought about by his technics: his technics which have solved some of his problems while also magnifying other problems.

SUMMARY: TECHNOLOGY'S CHANGES AND OUR DIMINISHING RELATIONSHIPS

Marcel questions, do technics principally improve or degrade human living? The short answer from this chapter: that depends on the situation. Marcel suggests reflecting on "the concrete relationship" between human persons and their technical processes.[145] Marcel's descriptions of the many changes brought about to human environments and communities sound similar to rich descriptions offered by media ecology scholars, Walter Ong, Lewis Mumford, and Jacques Ellul.[146] Media ecological approaches to studying relationships between human actions, technological inventions, and environments seek to avoid simplifying these interactions to being relations that *cause* or *determine* certain effects.[147] This work examines Marcel's reflections on the relationship between humans and their technologies. Strate notes that media ecology scholar Walter Ong describes this type of interaction as "relationism."[148] This summary of Marcel's account of technics' *impacts* on our environments, our lives, and our interpersonal interactions bears notes harmonious with media ecology.

Key media ecology scholars reflect on these relations between humans and their technologies by describing the *effects* of technics.[149] In reflecting on human interactions with their technologies, media ecology scholar Marshall McLuhan asks what does this "medium," (or technology), "enhance?" "obsolesce?" "retrieve?" "reverse?"[150] In a similar strain, Marcel notes that *the intention* of the inventor imbues the created technic for the improvement—or not—of the human person. Marcel's evaluation of the ethical dimensions in the creation of technics similar to Lewis Mumford's insights into technologies' impact on human histories. Mumford asks of the process of invention, although one *can* create a technology, *should* one create this technology?[151] Media ecology scholars offer questions that extract further reflection on *how* humans relate to their technologies. Neil Postman asked, "to what extent do new media enhance or diminish our moral sense, our capacity for goodness?"[152] Marcel asks different but important questions of the human experience of technology. Marcel offers similar evaluations of the ethical dimensions to our technologies "[T]echnology is good in itself . . . technology [could] be put to the wrong use."[153] Marcel calls for reflection on interaction between—the "concrete relationship" between—technologies and human beings.[154]

Marcel's insights on technics' ways of changing our human interactions provide important context for why persons may struggle to have any level of connection or communication with other persons. Our worlds in which we attempt to understand and communicate with other persons have been altered by technology at all levels and in all senses. Marcel identifies a diminishing of our feeling at home in our technical environments which produces feelings of fear or insecurity. These feelings of displacement from/in one's world become emotional and internal distractions to living in the present moment. Distractions while living the present moment diminish our ability to really be alongside/with the other person. Diminished ability to be *with* others will influence the quality of communication. Experiencing displacement from our *ecos,* we feel ill-at-ease in this context. Discord from one's environment presents barriers for communication: for intrapersonal and interpersonal communication.

Technics impact one's own internal world. In a world filled with technics, "he is increasingly incapable of controlling his technics, or rather of *controlling his own control*" and soon feels "at the mercy of his technics."[155] While our technics improve some areas of our control over our world, we also seem to lose control of our governing over these technics. Increasingly losing control over one's use of technics, one loses control of one's internal world, and one's intrapersonal understanding, autonomy, and communication.

Amid increasingly technical environments, persons may more frequently view themselves in relation to their machines and may adopt comparisons that lead to feeling less than one's technics or machines. In our comparisons to machines, we as human persons assume an incorrect method of evaluation of ourselves and of other persons. Measurements of efficient production are not accurate tools for measuring the quality or effectiveness of human beings. At times, the human person seems to produce very little and yet deserves to be valued as being irreplaceable and not to be ignored. Marcel's writings highlight the important problems in using measurements of technical efficiency for the human person. If we evaluate ourselves *as* human persons by using objective measurements, we miss deeper dimensions of being human that are not grasped by objective observations. A strictly technical evaluation of being human offers a diminished sense of full experience of being human. If one adopts a technical, objective approach to understanding one's own role as communicat*or,* one's understanding will be incomplete. One's diminished sense of the value of human life at all also leads to technical attitudes of devaluation of all human life.

Marcel invites a paused reflection on the "complicated relationship" between the technology-user and her technologies. Technology has altered how we view our world, ourselves, others, and our interactions with one another. At the heart of this alteration, Marcel identifies loss. To be human means to bear a sense of the sacred. Personally, and communally, persons have lost their sense that their life is sacred and that their life is a gift. With this lost sense of the sacredness of life, Marcel notes an increased self-centeredness. Centered more on oneself, persons are not attending to their relations with their world, themselves, and their neighbors. In this self-centeredness, persons withdraw from being present in their *ecos,* their dwellings. In this sole attention to one's own world, one loses sight of one's own life as situated alongside others. In thinking about one's own world, one loses sight of meeting ground for encountering others. Technics bringing displacement from *ecos,* diminished sense of one's life, and devaluation of the life of the other person presents human persons with problems for our communication. Displacements and disconnections alter the landscape for communicative encounters. Marcel suggests these alterations present us with more than a problem: our use of technology fills our horizon as a metaproblem—a problem that requires *more than* a simple solution. The next chapter, Chapter Three, seeks a diagnosis for this metaproblem, starting with reflection on our intrapersonal realm.

NOTES

1. Gabriel Marcel, *Man Against Mass Society*, translated by G. S. Fraser (Chicago: Henry Regnery Company, 2008), 62.

2. Gabriel Marcel, "Author's Preface to the English Edition," in *Metaphysical Journal*, translated by Bernard Wall, vii–xiii (Chicago: Henry Regnery Company, 1952), xiii. Marcel described his general approach to philosophy a "concrete approach" in which he reflected on particular aspects of life, drawing out reflections, and then applying his ideas to actual experiences. A concrete approach keeps the philosopher grounded in real experiences and applying ideas in actual life.

3. Dwayne A. Tunstall, *Doing Philosophy Personally: Thinking about Metaphysics, Theism, and Antiblack Racism* (New York: Fordham University Press, 2013), 66. In his discussion of technics, Marcel lacked empirical evidence for his general statements connecting techniques of degradation to dehumanization.

4. Plato, "The *Republic*, VIII" in *Philosophy of Technology: The Technological Condition: An Anthology*, 2nd ed., edited by Robert C. Scharff and Val Dusek, 9–18 (Malden, MA: John Wiley & Sons, Inc., 2014); Aristotle. "*Nicomachean Ethics*" in *Philosophy of Technology: The Technological Condition: An Anthology*, 2nd ed., edited by Robert C. Scharff and Val Dusek, 19–24 (Malden, MA: John Wiley & Sons, Inc., 2014).

5. Aristotle, "*Nicomachean Ethics*," 19, 20.

6. Ibid.

7. Robert C. Scharff and Val Dusek, "Introduction," in *Philosophy of technology: The Technological Condition: An Anthology*, 2nd edition, edited by Robert C. Scharff and Val Dusek, 3–7 (Malden, MA: John Wiley & Sons, Inc., 2014), 5. "For the Greeks, the cosmos is first of all φύσις (*physis*, from which our word 'physics' comes)—the whole of things, with all of its motion, changes of shape and size and physical development and growth, and generation and degeneration—and we are part of it, placed in it, and the human spirit thus seeks to understand it as that with which we are in any case involved."

8. Plato, *Gorgias*, translated by W.C. Helmbold (Boston: Pearson Education, 1952). Plato suggests that techne as craft knowledge, knowing how to, is not an art but a knack, while philosophy as episteme gets at knowing the true nature of things.

9. Robert C. Scharff and Val Dusek, "Introduction," 5. Scharff and Dusek compare ways of knowing in the classical Greek world to ways of knowing in the modern scientific world. Scholars within modern science view technological development in terms of progress and increased efficiency. One knows the world and gains power over creation.

10. Auguste Comte, "*Introduction to Positive Philosophy*" in *Philosophy of Technology: The Technological Condition: An Anthology*, 2nd ed., edited by Robert C. Scharff and Val Dusek, 54–67 (Malden, MA: John Wiley & Sons, Inc., 2014), 54.

11. Ibid.

12. Gabriel Marcel, *Being and Having: An Existentialist Diary*, translated by Katherine Farrer (New York: Harper and Row, 1965), 39.

13. Robert Rosthal, "Translator's Introduction," in *Creative Fidelity*, by Gabriel Marcel, ix–xxvi (New York: Farrar, Straus and Giroux Company, 1976).

14. Gabriel Marcel, *Metaphysical Journal*, translated by Bernard Wall (Chicago, IL: Henry Regnery Company, 1952), 97.

15. Gabriel Marcel, *The Mystery of Being: Reflection and Mystery*, vol. I, translated by G.S. Fraser (South Bend, IN: St. Augustine's Press, 2001), viii–ix. Marcel differentiated his description of transcendence from his contemporaries' definitions. For Marcel, transcendence did not signify realities impossible to access in the material world. One *could* experience transcendence within one's own inner world. A person strains toward transcendence and also may experience transcendent presences calling upon the person.

16. Marcel's descriptions of being and mystery are explained in Chapter Three of this project.

17. Gabriel Marcel, *Mystery of Being,* vol. I, 2.
18. Marcel, *Metaphysical Journal,* 97.
19. Plato, *Phaedrus,* translated by W.C. Helmbold and W. G. Rabinowitz (New York: Macmillan Publishing, 1956), 68–72.
20. Karl Marx, *"Capital: A Critique of Political Economy,"* in *Philosophy of Technology: The Technological Condition: An Anthology,* 2nd ed., edited by Robert C. Scharff and Val Dusek, 74–76. (Malden, MA: John Wiley & Sons, Inc., 2014). Modern philosopher Karl Marx while recognizing that science and technology were sometimes used to perpetuate class oppression also proposed that technology be used for good in establishing communist communities. Karl Marx and Friedrich Engels, *"The German Ideology"* in *Philosophy of Technology: The Technological Condition: An Anthology,* 2nd ed., edited by Robert C. Scharff and Val Dusek, 77–79 (Malden, MA: John Wiley & Sons, Inc., 2014). Marx and Engels connected technological advances with the evolution or progress of the human species—a suggestion that does not account for humans using technology as a tool of oppression.
21. Different twentieth-century scholars like Jacques Ellul, Herbert Marcuse, Marshall McLuhan, Neil Postman, Michel Foucault offer interpretations of how modern technology continues to improve or diminish the human condition, revealing or perpetuating underlying systems or dynamics of power, class, race, or gender. These studies of the role of technology in human life go beyond the limited scope of this discussion.
22. Martin Heidegger, *Being and Time,* 7th ed., translated by John Macquarrie and Edward Robinson (New York: Harper Collins, 1962). Marcel's thoughts on technology emerge concurrently with Martin Heidegger's philosophical interpretations of technology. In his 1927 publication *Being and Time,* Heidegger proposed that scientific, technological studies of being and of what is real were limited approaches for studying what it means to be in the world. In his discussions in *Being and Time,* Heidegger proposes that scientific approaches to studying being in the world do not lead one to understand being. Martin Heidegger, *The Question Concerning Technology and Other Essays,* translated by William Lovitt (New York: Harper Perennial Modern Thought, 2013). In this later reflection published in 1954, Heidegger changes his attention to this technology, inviting that we actually think about those relationships between people and things, between people and their technologies or tools. Although contemporary with Heidegger, Marcel does not make mention of being influenced by Heidegger's interpretations of how human persons engage technology.
23. Neil Postman, "The Reformed English Curriculum," in *High School 1980: The Shape of the Future in American Secondary Education,* edited by A. C. Eurich (Vintage Books, 1970), 161.
24. Ibid, 161.
25. Neil Postman, "The Humanism of Media Ecology," *Proceedings of the Media Ecology Association 1,* 10–16.
26. Lance Strate, *Media Ecology: An Approach to Understanding the Human Condition* (New York: Peter Lang, 2017).
27. Ibid., 2.
28. Lance Strate, *Echoes and Reflections: On Media Ecology as a Field of Study* (Cresskill, NJ: Hampton Press, 2006).
29. Strate, *Media Ecology,* 3.
30. Strate, *Media Ecology,* 10–11.
31. Donald F. Traub, *Toward a Fraternal Society: A Study of Gabriel Marcel's Approach to Being, Technology, and Intersubjectivity* (New York: Peter Lang, 1988). Thomas Anderson, "Technics and Atheism in Gabriel Marcel," *Bulletin de la Société Américaine de Philosophie de Langue Française,* vol. 7. no. 1/2 (1995), 59–68. Thomas Anderson, "The Body and Communities in Cyberspace: A Marcellian Analysis," *Ethics and Information Technology* vol. 2. no. 3 (2000), 153–158, 155. Thomas Anderson describes how Marcel's philosophy of "bodily co-presence" offers meaningful interpretations about the possibilities for and limitations to experiencing intersubjectivity in virtual communities.

Bernard Gendreau, "Gabriel Marcel's Personalist Ontological Approach to Technology," *The Personalist Forum*, vol. 15. no. 2. *Proceedings of the International Conference of Persons, Santa Fe, NM, 1999* (University of Illinois Press, 2003), 229–246. Philosophy scholar Gendreau summarizes Marcel's reflections that technology's dominant presence in human lives leads to multiple losses. Technology leads to the loss of authenticity, of integrity, of human creativity, of the true meaning of being human, and of an attraction to the sacred; and, technology contributes to a general loss of human reflection. Tunstall, *Doing Philosophy Personally*, 57–79. Tunstall in Chapter 3, "Living in a Broken World," summarizes Marcel's critique of the dehumanization brought by technology and Marcel's proposal for renewing the sense of sacred. Dennis Cali, "The Ecology of Presence and Intersubjectivity in the Philosophy of Gabriel Marcel." *China Media Research*, Vol. 11, no. 2 (2015), 109–117. Communication scholar Dennis Cali proposes that Marcel evaluates technology in how it affects experiences of presence between persons. Cali offers an insightful summary of some of the intersections in Marcel's description of presence and how technology impedes intersubjective encounters.

32. Lance Strate, *Echoes and Reflections.* Marcel is not identified as a philosopher reflecting on technology nor a scholar associated with the media ecology field.

33. In this note, I only offer dates for those technology scholars extensively addressed within media ecology. Martin Heidegger published initial insights into technology in 1927 with *Being and Time* and *in* 1954 with *The Question Concerning Technology.* From 1948 to 1951, Harold Innis, at the University of Toronto, published essays and articles on the role media has in history. Marshall McLuhan began publishing in 1951 with *The Mechanical Bride: Folklore of Industrial Man* and another significant work *The Gutenberg Galaxy: The Making of Typographic Man* in 1962. Lewis Mumford's *Technics and Civilization* was written in 1934. Jacques Ellul published *The Technological Society* in 1954 (a work translated into English in 1964), and into the 1980s. Walter Ong's *The Presence of the Word,* published in 1964, weighed in on the effects of technology on human communication.

34. Gabriel Marcel, "The Sacral in the Era of Technology," in *Searchings,* Gabriel Marcel, translated and edited by Wolfgan Ruf (New York: Newman Press, 1967). This lecture was published in *Searchings,* an edited volume of several of Marcel's lectures.

35. This lecture was published as the book *The Existential Background of Human Dignity* (Cambridge, MA: Harvard University Press, 1963).

36. Thomas Anderson, "Technics and atheism," 59.

37. For example, Wolfgang Ruf in his translation of Marcel's "The Sacral in the Era of Technology" used the word technology unless specifically using Marcel's words "technique" or "technical." *Tragic Wisdom and Beyond*, translated by Stephen Jolin and Peter McCormick, edited by John Wild (Evanston: Northwestern University Press, 1973). Translators Stephen Jolin and Peter McCormick in *Tragic Wisdom and Beyond* also used the word technology.

38. Gabriel Marcel, "On the Ontological Mystery" in *The Philosophy of Existentialism*, translated by Manya Harari, 9-46 (New York: Citadel Press Kensington Publishing Corp, 2002), 31. Manya Harari translated Marcel's French word as "technic." Gabriel Marcel, *Man Against Mass Society*. G.S. Fraser translator of *Man Against Mass Society* used the words "technique" and "technical" in Marcel's reflections on persons living in the modern world. Gabriel Marcel.

39. *Tragic Wisdom and Beyond,* 202. In that interview, Marcel stated, "the rejection of technology, of the world of techniques."

40. *Existential Background,*157-8.

41. Thomas Anderson, "Technics and Atheism," 59.

42. Marcel, "On the Ontological Mystery," 34.

43. Thomas Anderson, "Technics and Atheism," 59.

44. Marcel, "The Sacral" 43.

45. Ibid.

46. Marcel, *Man Against Mass Society,* 61.

47. Ibid., 9.

48. Howard P. Segal, "Introduction" in *Life in a Technocracy: What it Might Be Like,* by Harold Loeb, i–xxxviii (Syracuse, NY: Syracuse University Press, 1996). Marcel's particular description of technocracy sounds within a broader conversation about technocracy. The technocracy movement promoted industrialization and technological advances as a liberating force in modern society. Some of these ideas, taking from Henri Saint-Simon's proposals, were adapted in an American movement around technocracy that appeared in the U.S. during the 1920s into the 1930s. In a technocracy, government leaders should be engineers or be those who have practical knowledge and who organize society around the efficient production of and use of energy. Different leaders like Howard Scott and Harold Loeb within the technocracy movement offered variations on how a technocracy should function.

49. Marcel, *Man Against Mass Society,* 41.

50. Ibid., 153.

51. Marcel, *Man Against Mass Society,*195–196; *Existential Background,* 162.

52. Marcel, *Man Against Mass Society,* 30.

53. Ibid., 195.

54. Ibid., 195–196.

55. Neil Postman, "The Humanism of Media Ecology," 10.

56. Lance Strate, *Echoes and Reflections,* 17.

57. Lewis Mumford, *Technics and Civilization* (Chicago, IL: University of Chicago Press, 2010), *52.*

58. Ibid, 111.

59. Ibid, 321.

60. Ibid, 321.

61. Anderson, "Technics and Atheism," 59.

62. Jacques Ellul, "Understanding Our Age," in *Perspectives on Our Age Jacques Ellul Speaks on His Life and Work,* edited by Willem H. Vanderburg (Toronto, ON: House of Anansi Press, 2004), 26-7.

63. Ellul, "Understanding Our Age," 26–27.

64. Ellul, "On the Aims of a Philosophy of Technology" in *Philosophy of Technology: The Technological Condition: An Anthology,* 2nd ed., edited by Robert C. Scharff and Val Dusek (Malden, MA: John Wiley & Sons, Inc., 2014), 206.

65. Ibid., 206.

66. Ellul, "Understanding Our Age," 27.

67. Marcel, "The Sacral," 43.

68. Marcel, *Man Against Mass Society,* 61.

69. Nicholas Hewitt, *The Cambridge Companion to Modern French Culture* (Cambridge: Cambridge University Press, 2003).

70. John Herman Randall, *The Making of the Modern Mind,* revised edition (New York: Columbia University Press, 1976).

71. Stephen Schloesser, *Jazz Age Catholicism: Mystic Modernism in Postwar Paris, 1919–1933* (Toronto: University of Toronto Press, 2005).

72. Henri Bergson qtd. in Stephen Schloesser, *Jazz Age Catholicism,*10.

73. Marcel, "On the Ontological Mystery," 34. Marcel uses the phrase "withdrawal in recollection," as creators creatively produce the technic.

74. Marcel, "On the Ontological Mystery," 34.

75. Marcel, *Tragic Wisdom and Beyond,* 154.

76. Marcel, *Man Against Mass Society,* 41–42.

77. Ibid., 63–64.

78. Marcel, *Tragic Wisdom and Beyond,* 246.

79. Marcel, *Man Against Mass Society,* 62.

80. Ibid. Marcel's "concrete approach" attends to actual, real experiences of how human persons relate with technics.

81. Georges Friedmann, "The Social Consequences of Technical Progress," *International Social Science Bulletin: Social implications of Technical Change, UNESCO,* vol. 4, no.

2 (1952). Marcel referenced Georges Friedmann's descriptions of technical and natural environments. Friedmann's explanation of these environments appears in his 1952 article.

82. Gabriel Marcel, *The Decline of Wisdom*, translated by Manya Harari (New York: Philosophical Library, Inc., 1955), 3.

83. Marcel, *Existential Background*, 162. Marcel uses phrase "integrity of the human being."

84. Gabriel Marcel, *Homo Viator: Introduction to a Metaphysic of Hope*, translated by Emma Crawford (New York: Harper & Row, 1962), 80.

85. Ibid, 80.

86. Marcel, *Man Against Mass Society*, 63.

87. Ibid.

88. Marcel, *The Decline of Wisdom*, 17.

89. Marcel, *Man Against Mass Society*, 69. *The Decline of Wisdom*, 17.

90. Marcel, *Man Against Mass Society*, 14. *The Decline of Wisdom*, 17.

91. Marcel, *Homo Viator*, 82.

92. Ibid, 81.

93. Marcel, *Man Against Mass Society*, 74. "On the Ontological Mystery," 30.

94. Marcel, *Being and Having*, 74.

95. Marcel, "On the Ontological Mystery," 30.

96. Marcel, *Man Against Mass Society*, 44.

97. Ibid. Marcel states that in this way, one is primarily focused on "conserving the individual's own skin."

98. Marcel, *Man Against Mass Society*, 44.

99. Marcel, *Existential Background*, 160.

100. Marcel, *Man Against Mass Society*, 41.

101. Marcel's book *Being and Having* addresses these relationships to things and to one's body.

102. Marcel, *Being and Having*, 165.

103. Ibid, 164–165. Marcel uses the phrase, "they [things] reach me, one might say, underground," the things "tend to blot me out, although it is I who possess them."

104. Ibid, 164.

105. Marcel, "On the Ontological Mystery," 31. (Marcel's emphasis)

106. Marcel, *Mystery of Being*, vol. I, 25.

107. Marcel, *Man Against Mass Society*, 63.

108. Marcel, "On the Ontological Mystery," 33–34.

109. Virgil Gheorgiou, qtd. in Marcel, *Man Against Mass Society*, 173. In describing persons in a mechanized environment, Marcel quoted Virgil Gheorgiou's 1949 novel, *The 25th Hour*, to describe how people relate to others as though they are machines. Gheorgiou wrote: "The West has created a society which resembles a machine. It forces men to live in the heart of this society and to adapt themselves to the laws of the machine."

110. Marcel, *Existential Background*, 160.

111. Marcel, *Man Against Mass Society*, 40.

112. Marcel, "On the Ontological Mystery," 44. This was explored in Chapter One. Marcel wrote, "But what can be the value of such a philosophy for those who are a-Christian—for those who ignore Christianity or do not accept it? I would answer: it is quite possible that the existence of the fundamental Christian data may be necessary *in fact* to enable the mind to conceive some of the notions which I attempted to analyze; but these notions cannot be said to depend on the data of Christianity, and *they do not presuppose it*." Those who do not hold Marcel's same religious beliefs can still glean meaning from his descriptions of being, spirit, and life.

113. Marcel, *Man Against Mass Society*, 14.

114. Walter Ong, *The Presence of the Word: Some Prolegomena for Cultural and Religious History* (New Haven: Yale Univ. Press, 1967). Another Media Ecology scholar, Walter J. Ong, explores the meaning and experience of human presence. In *The Presence of the*

Word, Ong connects his discussion of human consciousness of presence to Martin Buber's description of voice, listening, and mysticism in the Hebrew writings. Thomas J. Farrell, *Walter Ong's Contributions to Cultural Studies: The Phenomenology of the Word and I-Thou Communication* (Creskill, NJ: Hampton Press Inc., 2000), 124–128. Ong reflects on changed experiences of presence amid cultural shifts in orality and literacy. Marcel's connection of awareness—or lack of awareness of—presence and of mystery to cultural changes in society concurs with Ong's reflections. Ong's works and Farrell's thorough study of Ong's works provide further exploration of how experiences of presence have changed in our technological world. I briefly revisit Ong's insights on presence in Chapter Five of this work.

115. Marcel, *Man Against Mass Society,* 55.

116. Marcel, *Tragic Wisdom and Beyond,* 155.

117. Marcel, *Tragic Wisdom and Beyond,* 247. Ricœur mentions in his interview of Gabriel Marcel. Marcel agrees with Ricœur's description of his own work and adds, "what might happen to the sacred" in a world of technology.

118. Gabriel Marcel, *The Mystery of Being: Faith and Reality,* vol. II, translated by René Hague (Chicago: Henry Regnery Company, 1951)

119. Ibid, 152.

120. Marcel, *Man Against Mass Society,* 40–44.

121. Ibid, 167–168. Marcel reflects on connections between the technocratic attitude and abstraction of multitudes of individual persons to a numbered mass. If one sees groups as faceless masses, there is a loss of "eschatological consciousness: the consciousness of the last things." Meaning that persons face individually where they are headed at the end of their biological life "the last things." Marcel identifies individual persons in concentration camps who testify to the value of the other's life by being willing to lay down their own life for another. Marcel suggests that individual witnesses to the value of this life point to a belief in life after death.

122. Marcel, *Mystery of Being,* vol. II, 148.

123. Ibid.

124. Marcel, "The Sacral," 51.

125. Marcel, *Being and Having,* 187.

126. Marcel, *Tragic Wisdom and Beyond,* 110.

127. Marcel, *Tragic Wisdom and Beyond,* 118. Oxford English Dictionary, "mirabilia, n," *OED Online.* Oxford University Press

128. Marcel, *Tragic Wisdom and Beyond,* 105–119. Marcel explores the meaning of "Every living thing is sacred" in his chapter, "Life and the Sacred."

129. Tunstall, *Doing Philosophy Personally,* 63–66. Tunstall summarizes Marcel's approach to technology as: "technology can degrade us by robbing us of our sense of the sacred" in our world and in life itself Tunstall describes Marcel's overarching approach to technology in his chapter named "Living in a Broken World" and what Marcel means by a loss of the sacred.

130. Marcel, *Man Against Mass Society,*70.

131. Marcel, *Creative Fidelity,* 54.

132. Ibid.

133. Marcel, "On the Ontological Mystery," 34. Marcel describes a feeling connected to one's technic use, a feeling that is a movement, a "stiffening, the contraction, the falling back on the self" and "self-centring and self-hypnotism."

134. Ibid., 40.

135. Marcel, *Man Against Mass Society,* 17.

136. Ibid., 1.

137. John Herman Randall, *The Making of the Modern Mind.*

138. Marcel, *Man Against Mass Society,*153. *Existential Background,* 123.

139. Marcel, *Mystery of Being,* vol. II, 148.

140. Marcel, *Mystery of Being,* vol. I, 29–30.

141. Marcel, *Man Against Mass Society,*14.

142. Ibid, 174. Marcel acknowledged that the concept "persons" and "personal rights" had been used by people who wished to justify and perpetuate class differences. Although the term "person" had been misused by some people to justify prejudices, Marcel argued that one could continue to legitimately use the term "person" in one's philosophical reflections.
143. Marcel, *Man Against Mass Society,*10.
144. Marcel, *Man Against Mass Society,* 10. *Mystery of Being,* vol. II, 148. Marcel gives these examples in both sources. Tunstall, *Doing Philosophy Personally,* 108. Tunstall notes how Marcel fails to address antiblack racism as an evident example of depersonalization that happens in Marcel's contemporary world. Marcel highlight experiences among white French persons of depersonalization during World War II, but "never addresses a major source of depersonalization in Western modernity, namely antiblack racism."
145. Marcel, *Man Against Mass Society,* 62.
146. Walter Ong, *Orality and Literacy: The Technologizing of the Word* (New York: Methuen, 1982). Mumford, *Technics and Civilization.* Ellul, "Understanding Our Age," 28. A thorough comparison of Marcel's characterization of technics' changes for human persons with media ecology literatures offers ground for future research. Marcel's descriptions could be brought alongside Ong's study of orality and literacy. Mumford's ages of technological ecologies like the "ecotechnic," "paleotechnic," or "neotechnic" phases. Ellul's descriptions of historical periods use an "all-inclusive" framework of *technique.*
147. Strate, *Media Ecology,* 154.
148. Ibid., 158.
149. Strate, *Media Ecology,* 221.
150. Marshall McLuhan and Eric McLuhan, *Laws of Media (*Toronto, CN: University of Toronto Press, 1988), 153. Strate, *Media Ecology,* 221.
151. Lewis Mumford, "The Technological Imperative," in *The Meaning of Technology: Selected Readings from American Source,* edited by Montserrat Ginés, 129–131 (Barcelona, Spain: Centre de Publicaciones del Campus Nord, 2003).
152. Neil Postman, "The Humanism of Media Ecology." *Proceedings of the Media Ecology Association 1,* 10–16. https://www.media-ecology.org/resources/Documents/Proceedings/v1/v1-02-Postman.pdf, 15.
153. Marcel, *Tragic Wisdom and Beyond,* 246.
154. Marcel, *Man Against Mass Society,* 62. Marcel's "concrete approach" attends to actual, real experiences of how human persons relate with technics.
155. Marcel, "On the Ontological Mystery," 31 (Marcel's emphasis).

THREE

Restlessness and Disunion Within

Our Problems with Being

Within the narrative of humanity only ever progressing throughout history, we witness ongoing hope that technology has, does, and will solve all problems that we face. In the 1930s, Marcel wrote, "[people] ask 'what can man achieve?' We continue to reply: He can achieve as much as his technics."[1] This attitude reveals an early-twentieth-century optimism rooted in what technics might help us achieve. Technology promised (and promises) to improve our living conditions, to cure illnesses, and to connect people. Yes, our technics have assisted us in improving the human condition, allowing more people to live a better, healthier, freer lives and increased our connections. But, alongside these improvements and technological advances, people have also reported increased experiences of fear for, uncertainty about, or loss of control of one's own life. In his same 1933 reflection on the impact of technics on our lives, Marcel described a sense that technics, while offering "partial triumphs," more evidently suggests "the failure of technics *as a whole*."[2] Persons who found technology's overall contributions to human experience to be negative, also experienced despair, this "despair consists in the recognition of the ultimate inefficacy of all technics."[3] People had hoped in technology and were seeing that their hope seems to have been misplaced. Marcel describes feelings of widespread hopelessness and uncertainty so markedly present in his 1920s and 1930s France, he called his environs an "era of despair."[4]

In Marcel's reflective approach to human experience, he paid attention to affective dimensions of being in the world. Marcel moves between noting communal manifestations of and personal feelings of despair and hopelessness. He seeks understanding of problems beneath the surface, what communication scholars call the realm of the intrapersonal communication. Feelings about life emerge in the context of communal approaches to life and life's meaning. Feelings about life also spring up from intrapersonal exploration and understanding. Marcel highlights the importance of one's evaluation of one's life. When this life loses its meaning, this life also loses its value.[5]

Marcel considers this loss of value for life to be the root of our problems in being in our technical modern world; and yet, people continue to put their trust first in technics to solve life's problems. When we are faced with a societal crisis such as terrorist violence or cybersecurity breaches, we conduct scientific studies to track who has access to weapons or to private information. These scientific studies provide key information to understand these problems, and yet fall short of helping us understand the crux or heart of these problems: the human persons who create or instigate these problems. Many contemporary experiences of our technological world are marked by displacements and disconnections, technological changes impacting human persons living in the world. Marcel suggests that what accompanies these changes brought on by technology is a "great impoverishment of our inner lives."[6] Because Marcel's reflection on technology begins and ends as a study of *human experience* of this technology, this work explores key vectors of human experience that impact and underlie our problems of intrapersonal and interpersonal communication.

Marcel connects the communal problems with personal problems along the vector of affect. We feel like our communal life is broken, we feel like our societal living is broken, we feel like our internal world is broken. Marcel exposes the roots of these feelings that something is broken in the heart of existence as feelings that something is broken in how I am in the world. Our problems of being in our technical world *begin* as problems of being at an intrapersonal level.

This chapter explores the question, what is at the heart of our problems with being in the broken world? This search begins with identifying common feelings about life that indicate our problems with living. Next, Marcel's distinctions between problem, mystery, and metaproblem set the stage for reflecting on experiences of a broken world. Then, this chapter identifies why technological problems require a metatechnological response. Finally, Marcel's suggested reflection on being signals *how* we might engage his metatechnological response. Problems in the broken world are rooted in problems of being. Marcel's philosophy continually calls for more than reflection. Digging deeper into the experience of being in a broken world, what will we uncover at the heart of these problems?

FEELING LIKE LIFE ITSELF IS BROKEN

Marcel notes how feelings of uneasiness and fear that are pervasive in people in society reveal deep human problems. Marcel, like preceding philosophers of existence Friedrich Nietzsche and Søren Kierkegaard, attended to feelings of dissatisfaction, anxiety, or anguish felt toward the problems in life.[7] One who studies existence attends to how "the union of *facticity* and *emotivity* begets the idea and the feeling of existence."[8] Existence is explored as being more than thought: existence involves participation, involvement, and feelings about this phenomenon or about life itself. Marcel joins his contemporary philosophers of existence when he describes these feelings about life that are felt personally and as a society. His metaphor "the broken world" describes how people *feel* amid modern society, a society marked by technical changes. People in a broken world experience restlessness, fears, and insecurities about life in general and their life in particular.

Marcel's study of the Parisian society uncovered persons who felt restless in their work and lives. Marcel wondered if this uneasiness pointed to inner disillusion with the world itself. If one is just what one has or what one's job is, one can feel "submerged by [one's] function" and settle into a "dull, intolerable unease."[9] Modern workers who feels reduced to their function experience the accompanying feeling of being uneasy or unsettled. Marcel describes another similar feeling about a life that is accelerated. In his 1933 play "The Broken World," the main character Christiane, while living a "busy, rushing life that she seems so much at home in, obviously masks an inner grief, an anguish."[10] Externally, modern individuals experience fast lives and much change due to societal and technological changes. Persons in this modern world might, like the character Christiane, mask inner experiences of emptiness. Marcel quoted Christiane's revealing words spoken at a key moment of self-revelation in the play:

> Don't you feel sometimes that we are living . . . if you can call it living . . . in a broken world? Yes, broken like a broken watch. The mainspring has stopped working. Just to look at it, nothing has changed. Everything is in place. But, put the watch to your ear, and you don't hear it ticking.[11]

Christiane describes her life unfolding in a world that she feels is not working nor is healthy. She qualifies the word "living" with "if you can call it living." The meaning of living life has been diminished and Christiane feels empty. In another reflection, Marcel notes that the modern person feels that life has "become blunted, and flat, and stale" and "whatever happens, it's all one to me, I couldn't . . . care less."[12] Marcel described visible sadness in the faces of many people he met at public and private gatherings.[13] As Marcel reflects on the visages of people around him, he finds loss, emptiness, and lack of passion for life. Marcel asked the questions, why do many people feel that life is missing something and what do people think is missing?

When one feels reduced to producer at work, one's insecurities about the value of one's own life also emerge. The worker who competes with machines in production, feels less productive and less valuable than the machine.[14] Workers fear loss of their jobs to the more productive machines. Along with fears for being able to sustain oneself in a job, one fears even more for one's children and their future. The cost of and "wear and tear" of caring for a child in this world becomes a worry for the parents, an added uncertainty amid the modern living.[15] Amid a world predicting future expansion of existing problems, Marcel noted the spreading sense of fear about and for human communities. Marcel asks, why do so many people today bear fears about life now and in the future?

Marcel witnessed people applying their own diminished view of the value of a life to the lives of the others around them. One looks at the other's life as productive, valuable, or serving a purpose in society. This technological view of the human person reduces the actual, concrete human person into an abstract entity of productivity or function. Marcel finds this "spirit of abstraction" applied and promoted throughout technological realms and workplaces: one becomes reduced to one's productivity.[16] The technological environment is structured to identify human persons according to productivity or contributions to the system. In this "spirit of abstraction" individuals become depersonalized and then can be treated as numbers, units that seem to be less-than human.[17] It is easier to treat an individual who is identified by one's number as just a number. Vincent P. Miceli suggests that these broken world interpersonal relations are marked by "divisiveness" that unsettles and puts communities at risk.[18] Marcel suggests that this temptation to treat human persons as less-than human is built into a world committed to technological success. Marcel asks, in our commitment to technological advances, why do we ignore the invaluable worth of each human person who is present in the situation?

These fears, uncertainties, and uneasiness about one's own life or the life of others may result in despair about life itself. When one reflects on one's life and feels one's life or quality of life is deeply "inadequate to something which I carry within me, which in a sense I am," one feels despair.[19] Marcel points to human persons *actually* bearing an innate sense that "I am." When one looks around at one's life and feels as if one is just performing functions or competing with a machine, one feels a vast disconnect between one's external functioning and one's innate sense that one *is*.[20] Marcel suggests "life in a world centered on function is liable to despair because in reality, this world is empty, it rings hollow."[21] This world of technological functions rings hollow in comparison to what one feels in one's heart, this sense: "I am." If one tries to find one's meaning in one's function in or one's level of production for society, Marcel suggests one will find no sign of lasting meaning. If one evaluates the world and one's role in it and finds nothing of lasting value, one's despair at finding no thing of value may drive one to suicide. Marcel notes suicide is always a possible response to an evaluation of life itself.[22] Fears, insecurities, and uncertainties about life may lead to this evaluation: my life no longer should be lived. Marcel noted the world we live in "may even seem to counsel—absolute despair."[23] Fears, insecurities, and despair have become common feelings about life itself.

In Marcel's identification of problems, we uncover descriptions reminiscent of the philosophies of media ecology scholars. Susanne Langer notes how "a philosophy is characterized more by the *formulation* of its problems than by its solution of them" and "this disposition of problems is the most important thing that a school, a movement or an age contributes."[24] Marcel offers insightful formulations of the problems he felt our broken world faced, and in his articulation of the problems in relations in a technological society, Marcel's philosophy extends and contributes to media ecologists' questions.

Marcel connects these uncertainties, fears, and despair to the deepest human questions of and about life. What are we missing about being human when we rush and reduce a person to function? What do people think it means to live life today? Why do so many people today bear fears about life now and in the future? These deepest human questions of one's own life enunciate inner dialogues at the heart of human experience. Questions echoing as intrapersonal dialogues reveal personal attitudes toward communion.

Marcel's questions present themselves to us as both theoretical and technical problems.[25] Different persons and philosophies offer varying answers to these and all human questions. These questions uncover our communal problems: we have different meanings for what life is, for what life is for, or what a person is. Questions we ask ourselves uncover problems in understanding ourselves. This project pauses to identify how persons might explore these questions. As noted in Chapter One, these questions are rooted in the question, "Who am I—I who question being?"[26] Intrapersonal communication deals with how one understands who one is. The question "Who am I" is more than a problem requiring more than a simple answer.

ARE WE FACING A PROBLEM OR A MYSTERY?

Marcel suggests that modern persons propose to answer questions like "Who am I—I who question being?" from an objective scientific perspective.[27] In a scientific schema, one stands back from these questions about life and offers a rational answer to the questions. Making predictions, modern science promised and continues to promise technological advances will assist us in having increased control over nature, over the world around us.[28] Technical advances promised greater human efficiency for workers, and yet also brought displacement of human workers by machines. Marcel identifies a fundamental problem with this proposal that one can take an objective, scientific approach to study a human problem about who one is.

Marcel distinguishes among our human problems and questions as different types of situations. He makes an "essential distinction" between problem and mystery.[29] In reflecting on certain questions or problems, one is oriented toward the question as if one stands before the idea. Marcel calls this particular situation a problem. One is able to assume an objective approach as outsider looking at the problem. Before a problem: "I, who formulate this problem, should be able to remain *outside* of it—*before* or *beyond* the problem that I state."[30] He explained that "a problem is something which I meet, which I find complete before me, but which I can therefore lay siege to and reduce."[31] One may use an objective, scientific method to think about and solve a problem about mechanical speed or efficiency.[32] An objective stance allows one to assume a distance from the situation to get an uninvolved analysis of the situation.

Yet, some of the problematic situations that need solving deal with human experience and one cannot get an "objective" distanced perspective to this sort of problem. Marcel calls this kind of situation a mystery. A mystery is "a problem which encroaches upon its own data, that invades the data and thereby transcends itself as a simple problem."[33] One cannot position oneself outside the mystery to assume an objective perspective. A mystery "is something in which I find myself involved, whose essence therefore, is not to be completely before me."[34] One studies a mystery in a different way than one studies a problem. When one reflects upon a question or situation that is a mystery, one cannot approach this reflection using an objective, scientific method.[35] The one who reflects upon a mystery is involved in and acted upon by the situation. An example of an experience of mystery is how long work hours affect the human person. If one wants to reflect upon one's bodily experience, one cannot separate oneself from one's body in order to make an abstraction of the selfsame. One is involved in the problem, the question touching on something which is "not completely before me," so one cannot reflect on this situation in the same way.

Marcel identifies situations that touch on the human person as mysteries. Some examples of mysteries he describes include the mystery of body, relationships with other persons, and the mystery of being itself. This distinction between mystery and problem helps us understand being itself. The difference between problem and mystery came to him suddenly one day and would remain a significant theme throughout his life.[36] He differentiates a "mystery of being" from a "problem of being."[37] When being is studied as a problem, one loses an element of how the one who studies it is in actuality involved in the problem.

A mystery is *more than* a problem, involving the one who reflects and the one who is reflected upon. Marcel sometimes used the word meta-problematic to describe a problem that had "metamorphosed to mystery."[38] The problem has become more-than-a-problem. As one attempts to study a problem about one's own life, one encounters "a peculiar opacity," a realm beyond resolve, a metaproblem.[39] When one is reflecting upon one's own being or mystery, one has "advance[d] into the realm of the metaproblematic."[40] Because one's own being cannot itself be an "object of thought," the action of participating-in-being is meta-problematical.[41] Dwayne A. Tunstall uses Marcel's term metaproblem for a situation in which one simultaneously identifies being and recognizes that being over-abounds the subject.[42] The one who studies being cannot ever get "outside of" or at a distance from being so as to study it. Marcel notes the issue with using spatial descriptors for explaining *how* we think about phenomenon like being. We experience being in a way that we can stand "before" the other who is with me, and yet also experience my own idea of the other who is before me, as some reality that I am aware of "within" myself. The spatial categories "before" or "within" cannot adequately describe one's experience of being, "these categories are transcended."[43] In the broken world, the reflecting subject encounters situations in her own existence that go beyond an objective study. She finds herself in the presence of a realm beyond problems, in a metaproblematic realm.

OUR PROBLEMS WITH TECHNOLOGY REQUIRE A METATECHNOLOGICAL RESPONSE

Technics promised to aid in solving human problems, yet those problems have not disappeared. Technics help us find solutions for certain problems but do not offer solutions for these human problems of fear, uncertainties, or despair. In our technological advances, if we have not progressed past fear or despair, is this actually progress?[44] Marcel suggests the proliferation of fears indicates certain human problems cannot be solved with an objective problem-solving method, these problems have become metaproblems.

Marcel gave a particular descriptor to our metaproblems associated with technology: "metatechnological problems."[45] Metatechnological problems range from personal lack of control over one's own technics to international struggles for ownership of the most advanced technologies. On a personal level, one finds oneself "at the mercy of his technics, he could not control technology, nor control his own control of technology."[46] This lack of control of technology permeates society from the intrapersonal level up to the global level at which nations and corporations interact. Although modern governments and scientists continue to promise increased control over nature itself—as the earth and the environment—these same people demonstrate a "powerlessness and confusion" in having any control over "the destiny of our planet."[47] Marcel offers the 1960s space race as one example of a metaproblem associated with technology. The technological goals assumed by national superpowers to conquer space involved spending vast amounts of money and engineering to reach distances beyond earth. Marcel describes this space race of revealing an "unrestrained colonialism" that focuses on increasing technological might and not on solving problems among men and women on earth.[48]

Experts continue to propose solutions for worldwide problems that address only the material dimensions of these problems.[49] Strictly scientific thinking suggests our "world can be controlled and mastered by technological thinking."[50] Many communities and governments put their hopes in increased industrialization to improve lives with increased planning and organizing of our cities and countries.[51] Although technological advances promise progress and improvement, "crises, catastrophes, even natural upheavals" continue to disrupt our lives.[52] People still hold on to "an absurd illusion [thinking] that by means of some sort of world planning these calamities can be kept permanently at bay."[53] The positive aspects of technological planning are not to be dismissed.[54] However, the breadth or scale of these problems cannot be addressed as mere problems. Our community, national, and world problems appear on a vast scale.[55] Marcel gives these types of mega-problems a new category: a metaproblem. Our metaproblems require *more than* a technological response.

Marcel called for a "metatechnical" response to the metaproblems of technics. A metatechnical response "involves a critical questioning of technology, of the very notion of technique."[56] Critical questioning involves a standing back to question the premises and foundations of technology. Critical questioning asks probing questions of our broken world- - What are these metaproblems due to? What or who is at the root of these metaproblems? Reflective questioning may lead to creative human responses.

Technology has introduced metaproblems because this technological progress has happened without simultaneous attention to the effects on human persons. Marcel identifies that we have long ignored how the development of war technology, systematization of workplaces, and unchecked technological expansion has affected the human persons involved in these processes.[57] One experiences the "burden of technics" and being "dragged down into the excesses of technocracy."[58] Technology is felt as burden upon human persons. We cannot "return to a pre-technical age," but are invited to "struggle efficaciously against the weight by which in a technical age [one] is dragged down into the excesses of technocracy."[59] In our technological advances, we have ignored the many ways technology has negatively impacted the human person. Sometimes communities seem motivated by the "technological imperative" that claims because we *can* invent some tool or technology we *should*.[60]

According to Marcel, scholars who reflect on the world should call attention to this devastating neglect of the human person. He claims, "philosophy must bring to light the profound but usually unarticulated uneasiness man experiences in this technocratic, bureaucratic milieu where what is deepest in him is not only ignored but trampled underfoot."[61] Assuming a metatechnical lens, philosophy should attend to the actual effects of technological advances on human persons: on the presence of fears, anxieties, uncertainties, or despair in their experience.

Marcel connects some of these metaproblems in technology with nihilistic interpretations of life itself. Marcel differentiates his view of human life from his contemporary Jean-Paul Sartre's perspective on human life. Marcel suggests that Sartre and Martin Heidegger describe the human person "as the victim of some cosmic catastrophe, flung into an alien universe to which [one] is bound by nothing."[62] With this view of the human person, one is "unable to recognize life is a gift."[63] In a world where technology is prioritized, the human person is treated as a machine, a compilation of parts, productive or unproductive. Marcel suggested that "technocracy is inevitably nihilist."[64] In a technocracy, one evaluates human living as coming from and leading toward nothing.

Marcel invited interpreting the broken world with a lens that was not nihilist. Yes, one can interpret life in this world as being meaningless, but one can also "refus[e] to accept an intolerable situation as final."[65] Marcel calls this refusal hope. Hope involves opening to believe in persons in the world broken by technology who *may heal*. For Marcel, "hope transcends imagination" and "transcends all laying down of conditions."[66] In hope, one thinks about one's situation from an *other* perspective. One identifies nihilism as a way to see life and one refuses this interpretation. One hopes in what life *might* mean. A metatechnical approach involves refusing a nihilistic interpretation and seeking hope in life's meaning and in the person's capacity to heal.

One of Marcel's major issues with our technological world is the dehumanization that occurs. If one employs a spirit of abstraction in dealing with the human person, one ends up "degrading" the person. If one views the human person only as a material entity who lives and works in the world as a material body, one degrades who the human person is to something less than human.[67]

Marcel's metatechnical approach uncovers how metaproblems in the world broken by technology emerge when we ignore effects on the human person, interpret human life as being meaningless, and dehumanize the persons who dwell in society. The roots of the problems in our technological world spring from a lack of awareness that *being* shapes the heart of human existence. According to Marcel, persons misunderstand the meaning and value of their own lives. Marcel described how persons in the broken world see life,

> The great majority of human beings grope about during their whole lives among these data of their own existence rather as one gropes one's way between heavy chairs and tables in a darkened room. And what is tragic about their condition is that perhaps only because their lives are passed in this shadowy gloom can they bear to live at all. It is just as if their seeing apparatus had become finally adapted to this twilight state.[68]

Persons in the broken world try to make sense of the "data of their own existence." Yet, they work in a darkened state unable to see or make sense of the world around them. The persons possess "seeing apparatus" that have adapted to a twilight state, to an uncertain groping one's way through situations.

METATECHNOLOGICAL SOLUTIONS: AWAKENING TO BEING

Marcel's analysis of technology does not offer specific solutions for these metaproblems presented by technics. He proposes a metatechnical approach of critical questioning of the existing technological solutions for our human problems. Marcel invites considerations of factors, dimensions, interpretations, or persons who may have been overlooked. If persons stumble around in a "twilight state," Marcel notes ways of awakening from this state.

While Marcel does not offer specific protocol for addressing these problems, he does offer a philosophical approach—a way of reflecting upon and participating in—for navigating our metaproblems with technology in enlightening, life-giving, caring, and/or loving ways. Some scholars offer scientific, objective solutions for our metaproblems, Marcel suggests a metatechnological response that is a creative response: a reflection upon elements that go beyond initial reflections, technological assessments, or practical solutions. This work proposes that Marcel's philosophy of being—which includes reflecting upon being, body, and being *with* others—is a creative metatechnological response to the metaproblems of technology. Awakening to being offers a healing response to our problems of technology. Marcel's philosophy of being involves a particular way of reflecting for thinking about and contemplating human phenomena, phenomena that involve more than material, physical realities.

Philosophical Reflection

Philosophy offers a metatechnical approach, a way of approaching our problems that proliferate on personal, communal, and societal levels. Philosophy analyzes technological extremes experienced by humans as a result of totalitarian warfare, atomic bombs, or concentration camps. Technological possibilities and actions bring earth's inhabitants along a ledge of fears that humans will engage annihilating "solutions" to deal with territorial or economic problems. A philosophers' role is to uncover the *effects* of these technological programs on the human persons living in this world. According to Donald F. Traub, Marcel's call to philosophers is a vital call, "the reflective thinking of the philosopher must stand against the wretchedness of a world that could be destroyed by the techniques man has perfected."[69] Philosophers (should) speak to humanity about the human person who is irreducible and not to be commodified for any reason at all—for a technological, political, economic, or communal purpose. Marcel identifies ways philosophers could reflect on being human.

The last section of this chapter Three and the next two chapters in this work, Chapters Four, and Five, step aside from reflection on technology per se to attend in depth to what Marcel means by "impoverishment of our inner lives." The discussion of technology will be brought to the foreground again in Chapter Six, the final chapter of this work. Because Marcel's reflection on technology begins and ends as a study of human experience of this technology, this work explores in depth key vectors of human experience that impact and underlie our problems of intrapersonal and interpersonal communication.

For Marcel, philosophical thinking, which can be engaged by everyday persons, involves reflecting upon phenomena outside the person and also reflecting upon the *one* reflecting. He stated, "[T]he distinctive note of philosophic thought . . . is that not only does it move toward the object whose nature it seeks to discover, but at the same time it is alert for a certain music that arises from its own inner nature if it is succeeding in carrying out its task."[70] Reflection involves a two-part movement of understanding: an active search *active* search for understanding phenomena and *passive* encountering of understanding. Reflection means "attention to," looking and thinking about phenomena.[71]

Because these problems about technology deal with human persons and their relations in the world, the situation—more than a problem to be solved—is a mystery for reflection. Marcel distinguishes two types of reflection for considering problems versus mysteries. One uses primary reflection—"'*pensée pensée*,' or 'first reflection'"—for problems and secondary reflection—"'*pensée pensante*' or 'second reflection'"—for mysteries.[72]

One uses primary reflection when one reflects on problems, as one thinks analytically and is able to study the problem as an object.[73] One assumes this position of primary reflection as a "detached and technical thinking."[74] Primary reflection includes abstract thought that is carried out without being influenced/shaped by the thinking subject.[75] In primary reflection, one thinks abstractly and in this abstracting "reifies thought and gives it its own ontological status."[76] Marcel suggests that this way of thinking allows one to reflect upon problems that deal with physical, material, or measurable realities.

However, primary reflection does not allow one to understand meta-problems and mysteries. Realities like being, human experience, or body invite a distinct type of reflection. When one using primary reflection finds oneself "checked by a certain break in the continuity of experience," Marcel argued, "it becomes necessary to pass from one level to another through reflection.[77] Secondary reflection, or, involves "a movement of retrieval which consists in becoming aware of the partial and even suspect character of the purely analytical procedure [of primary reflection]."[78] Secondary reflection involves a standing back to become aware of limits within a strictly analytic primary reflection upon situations. Marcel called this movement of retrieval a returning to a broader scope for reflection, stepping back to attend to more aspects. To understand what Marcel means by secondary reflection, we must take a step back to understand how Marcel approaches reflecting on being.

Marcel's Approach to Reflecting on Being

Marcel suggested that a method for reflecting on being is built upon assumptions about philosophy itself. Marcel does not study being as categorized into metaphysical hierarchies or definitions.[79] Marcel's starting point for philosophy is a realization that one is always-already in relation to being. His foundational question, "Who am I—I who question being?" uncovers being as ground and foundation of all questioning and reflection.[80] In asking this question of existence, Marcel asked, "[H]ow am I qualified to begin this investigation?"[81] He did not agree with René Descartes' *cogito* as the starting point for philosophy, *cogito* being an "epistemological subject as organ of objective cognition."[82] Marcel proposed that an "I am" is the starting point for all reflection on reality, the "*I am* . . . presents itself as an inseparable whole."[83] Marcel asked both "Am I?" alongside his questions of "[W]ho am I, who question being?"[84] Marcel describes the person, as a being in a situation.[85] One can never think about being or being in the world from an objective, outsider position. The one reflecting on the question remains an important participant in reflection.

Marcel eschewed viewing the subject "I"—who reflects on what it means that "I am"—as a separate cognitive entity. As one is always a being in a situation—the first, foundational situation is being as this body.[86] A reflecting subject can never stand apart from her mind or her bodily reality.[87] Marcel did not see the mind as structuring reality. One could not disengage the "being who thinks" from the being "who endeavors to *think* himself."[88] As bodily reality, she also bears senses of and feelings about phenomena in the world. She is vital living being who thinks about the world. Thus, the one who reflects on life does so as mind and body. One relates intellectually and vitally to reality, the world, and being. Marcel explores the sense or feelings one bears about the world as a type of intuition.

For Marcel, thinking involved an element of coming into the presence of. He described thought as "deepening of metaphysical knowledge" which means experience "turns inwards towards the realization of itself."[89] Words like "turning inwards" and "deepening" signal actions like entering into or coming into the presence of. Understanding as a coming-into-the-presence-of signals a presence already there before the questioning. Marcel interpreted the already-there presence as *being*. One reflects on one's own life as a "totality of being" and one reflects "about [oneself] as a totality."[90] A totality of being is a realm in which one participates. Before one reflects, one *was*, and one *was in* the presence of being. Marcel acknowledged "that the whole reflexive process remains within a certain assertion which I *am*-—rather than *which I pronounce*—an assertion of which I am the place, and not the subject."[91] For Marcel, existence precedes reflection: one *is* before one *thinks.* I am and therefore, I think about *that* and *how* I am.

For Marcel, being over-abounds one's full reflective or objective analysis. Being's intimate connection to thinking prevents one from standing apart from being in order to define it. In his notes from 1912 to 1913, Marcel suggested that "being is neither substance nor representation. It can only be conceived as that in which thought participates."[92] Being holds intrinsic significance that "withstands—what would withstand—an exhaustive analysis bearing on the data of experience and aiming to reduce them step by step to elements increasingly devoid of intrinsic or significant value."[93] In this sense being, as the possibility of experiencing inexhaustible phenomena, is engaged by and over-abounds a person's thinking or description. The one reflecting upon one's being takes the simultaneous roles of reflecting subject and participant-in-being. Marcel used the term *participation* to describe how one thinks about being. For Marcel, participating in being involves both active and receptive dimensions.[94] When one tries to get to the meaning of one's own being, one never arrives at complete or exhaustive understanding of what one's being means. One reflects upon being, understanding some of what being is and remaining in the presence of being that over-abounds.

Marcel's approach to being would not be an ontology as a systematic description of being, but more a discovery about who we are. Dwayne A. Tunstall stated that Marcel reflected upon being as "who we are and not what we are."[95] Marcel's descriptions of phenomena like being are not definitional ontologies but "exploratory interpretations of the meaning and experiences of" those phenomena.[96] One participates in being and cannot fully comprehend what being means. Being eludes definitive description, and yet, being is real. Marcel sought to describe what it means to be in the world and the way one seeks for these descriptions is secondary reflection.

Secondary Reflection

One of Marcel's contributions to philosophy was his phrase mystery of being, the concept he explored in his two Gifford Lectures, both titled Mystery of Being, in 1949 and 1950. Marcel described this insight as an event that happened *to* him. Marcel suggested he uncovered his distinction between problem and mystery and the phrase "mystery of being" by surprise. He described the distinction "which suddenly forced itself upon me in the winter of 1932" "was revealed to me in a flash of insight, a *syneidesis*."[97] When I "recognize" mystery, "I found myself acting on an intuition which I possess without immediately knowing myself to possess it."[98] One's feeling about or sense of is an "intuition which cannot be, strictly speaking, self-conscious."[99] Marcel suggests all persons bear an intuition about being.

One bears a sense or feeling *before* one seeks clarity through reflection. Marcel also calls this sense or feeling, a "presentiment or forefeeling of something regarding which we can say, 'This is reality.'"[100] Although he promoted a concrete approach to reflection, he contended that all thinking starts from an "inaccessible" source.[101] One's "blind intuition," "'forefeeling' or premonition" forms one's ground for reflection on existence.[102] Before one reflects upon existence, one starts out from somewhere, some ground. Marcel named being as the starting ground for his examination of human existence. In his later writings, Marcel preferred the term "insight" instead of "blind intuition" to describe this source for one's thoughts about being. He clarified that this insight, or "*syneidesis*. . . a kind of vision which brings things together" is the "most adequate" description for the "flash of insight" about being, mystery, or the mystery of being.[103] One has a sense about the phenomenon and seeks clarity about that sense through *more* reflection.

Marcel calls this *more than* reflection a "reflection upon this reflection (in a reflection 'squared')."[104] This reflection upon reflection involves a detaching from experience in order to be able to understand these situations.[105] Marcel named this detaching from experience, "second reflection," "recollection."[106] He explicated recollection as involving abandonment:

> The act whereby I re-collect myself as a unity; but this hold, this grasp upon myself, is also relaxation and abandon. *Abandon* to . . . *relaxation in*

> *the presence of*. . .yet there is no noun for these prepositions to govern. The way stops at the threshold.[107]

Recollection involves a collecting oneself, a becoming present to oneself. Again, Marcel returned to an active-passive stance in selecting terms like "abandon to" or "relaxation in the presence of." In his text, Marcel did not specify the objects or nouns to which these verbs "abandon to" were resolved. He spoke vaguely of descriptive categories for these experiences. One "stops at the threshold" of complete descriptions. In recollection, one understands through *recognition* of who "I myself am as a unity," about what it means to be.

When reflecting upon a phenomena like being one recognizes what being means. This recognizing what being means is "intuition . . . which can grasp itself only through the modes of experience in which its image is reflected, and which it lights up by being thus reflected in them."[108] When recognizing what this phenomenon means or *is*, one's understanding about the phenomenon coincides with one's feeling or sense about the phenomena. Dwayne A. Tunstall suggested that a better way to describe what Marcel means by "intuition" is not as a visual perceiving of being but as an "appreciating being."[109] In perceiving something, one knows that thing as an object. In appreciating something, one comes into the presence of that thing.[110]

In secondary reflection, reflection upon reflection, the one who reflects is "consciousness seeking to be conscious of itself."[111] Recollection involves a person reflecting upon phenomena in life that cannot be understood through philosophical representation. One is reflecting upon being, life, mystery, metaproblems, all of which the one who reflects is deeply implicated and involved in the reflection. Within recollection, one becomes aware of one's power to position oneself so as to understand one's life and one's power to uncover what one's own position means. In this secondary reflection, "I ask myself how and from what starting point I was able to proceed in my initial reflection, which itself postulated the ontological."[112] To use secondary reflection to reflect upon what one's own life means, Marcel explains,

> Capable of taking up my position—in regard to my life; I withdraw from it in a certain way, but not as the pure subject of cognition, *in this withdrawal I carry with me that which I am and which perhaps life is not.* This brings out the gap between my being and my life. I am not my life; and if I can judge my life . . . it is only on condition that I encounter myself within recollection.[113]

Secondary reflection involves an act of withdrawal as a drawing back in a shadow/reflective response to life. Marcel described how one knows one's self or grasps the meaning of one's inner world. Recollection is more than a subject "looking at" phenomena as objects of cognition. In secondary reflection, as one reflects about one's own life, one is in position and takes up a position toward or within one's being-in-life.

One has a sense about life, being in the world, and in secondary reflection one reflects on *my* life, *my* being, *my* being in this situation. In secondary reflection, one seeks for descriptions of, analogies for, and the meaning of particular aspect of one's existence, one's situation, one's life. Marcel uses secondary reflection when thinking about body is always my body, (to be discussed in Chapter Four), and he demonstrates the exploratory process of reflecting, reflecting again, and probing the ideas for inconsistencies or insufficiencies. Marcel seeks for "categories" to describe human existence.[114] When one proposes a category for describing an aspect of life and finds no insufficiencies as a descriptor, one has a sense that one has uncovered a valid category for describing human existence. When Marcel proposed "mystery of being" as a category for human existence, he believed this a significant descriptor that opens up what it means to be human. One participates in the "mystery of being," participating in it and able only to identify that one "finds oneself in the presence of a mystery."[115] These categories for human existence offer the human person her most vital, most significant explanations for and descriptions of life itself.[116]

In his projects, Marcel does not begin first with logical deduction so as to uncover a set of truths about life. His philosophy of being is a "listening to the voices and appeals comprising that symphony of Being—which is for me, in the final analysis, a supra-rational unity beyond images, words, and concepts."[117] Marcel's philosophy does not begin as his own search but he *is called* into searching for meaning by "voices and appeals" that come from a source outside himself. He emphasizes, "it must be remembered that my thinking takes its departure above all from feeling, from reflection on feeling and on its implications."[118] Although Marcel's writing may imply that all people share a similar foundational forefeeling about being, mystery, or being as body, his writing remains open to our interpretations which approach being in the world from different perspectives.

Ontological Exigence

This chapter explores, what is at the heart of our problems in being in the broken world? Equipped with secondary reflection as a way of reflecting on our metaproblems created by technology, we turn again to uncover the source of our metaproblems with technologies. Marcel described experiences of objectification, reduction, and degradation of persons amid technological environments. The person sometimes feels degraded to a status of thing as if they had lost their status of being.[119] People living and working amid a technological society feel as if they are functionaries, and "we, at least some of us, feel acutely this lack of something, this impoverishment."[120] His experience the modern world around him—of "the tremendous event which laid waste or maimed our existence from the year 1914"—moved Marcel to search to understand the effects on the human person.[121] Following his recollection of the tragic experiences from the Great War, Marcel identifies "metaphysical uneasiness," "to be uncertain of one's centre, it is to be in search of one's own equilibrium."[122] In our technological society, the "system which [we] have adopted involves a tragic oversight of certain deep human exigencies."[123] Marcel turned his reflection upon this metaphysical uneasiness that he sensed lay at the heart of modern experiences.

In his reflection on the heart of his own experience, Marcel identified a sense of an "ontological exigence," a demand to recognize being as that which cannot be reduced to an abstract reality or thought away. If one tries to say that at the heart of existence is nothing, "being, grasped beneath the level of all objectification, is the fundamental and indissoluble bond."[124] *That* I am means being is *irreducible* and being is *how I am* existing at all. If one tries to make oneself an object, a functionary, a part of the machine, in reflecting on who one is, one meets with something irreducible at one's core. When one peels back the layers on who one is, one comes to an irreducible fact of existence: one is. Marcel interprets "being is given. . . as it is truly a *gift*" not a gift that is a thing but being "as an act."[125] Being is given. One is not only a subject or agent who *is* in some way, one receives being as gift. One cannot think away that which is given to one, for its source is not within one's own mind.

Because one is not the sole actor in, nor independent possessor of, being, being is experienced as demand. Marcel notes that some translators offered the phrase "ontological need" for his description of exigence. He clarified, being is "something that is *demanded*."[126] Ontological exigence, the exigence of being, is "a deep-rooted interior urge. . . an appeal."[127] In his earlier writing, he described this call of being, "*appeal qua appeal*, I am led to recognise that the appeal is possible only because deep down in me there is something other than me, something further within me than I am myself."[128] One cannot ignore the presence of nor silence the call of being.[129] Being as appeal *upon* demands acknowledgment. When reflecting on what one's life is worth, one encounters a value at one's deepest levels, my being, *that* I am, ontological exigence, being that demands acknowledgment.

At the heart of experience, one feels this longing to be with others, an emptiness that calls out *for* acknowledgment. One always has the possibility to ignore this feeling, to negate being, but this seems to be connected to widespread experiences of despair and dissatisfaction with life seen in a broken world.[130] The opposite of this negation of being would be to hope in being: to acknowledge that at the heart of one's own life one finds oneself in relation to others, one who is "permeable to" the presence of others.[131] Marcel suggests being aware of this presence is a hope that brings fulfillment. One finds fulfillment in one's work and life when one is conscious of satisfying "a profound requirement."[132] This feeling of fulfillment happens on a deep internal realm a "domain of either personal or inter-personal intersubjectivity."[133] For an individual human being, feelings of fulfillment in and about life are rooted in interior domains that are fundamentally intersubjective. Fulfilment in being alive stands in distinction from feeling dissatisfied with life. Reflecting on one's uneasiness, one may find the outline of ontological exigence. "Exigence only takes its meaning, its value as aspiration, in a being who is torn apart and suffering, or who is exiled and more and more painfully aware of this exile."[134] Ontological exigence is sensed when one feels most profoundly the *lack* of being. Marcel invites awakening to this ontological appeal at the heart of being: a calling that one always already is *with* others.

SUMMARY: RESTLESSNESS AND DISUNION WITHIN

An "I am" dwells at the heart of *how* we are or communicate *with* others. An I who understands oneself can engage intrapersonal communication with oneself. Gabriel Marcel invites a thorough self-reflection marked by philosophical reflection and attention to being. This work proposes that Marcel's tool for philosophical reflection as secondary reflection offers a creative way for learning intrapersonal dialogue. Before imagining openings for dialogue with another person, one must understand (to some extent)[135] who one is as dialoguer.

This chapter began by identifying common feelings of restlessness, fear, and insecurity about life itself. Many persons feel that life itself is broken. Marcel suggests that these feelings about life uncover our deepest human questions about that life, one of which is, what is at the heart of our problems in being in the broken world? These feelings for and questions of one's own life enunciate inner dialogues at the heart of human experience. These feelings indicate that underlying human problems seem to be more than simple solvable issues. Marcel suggests that we approach these feelings and questions as mysteries, as more than problems, as metaproblems. Thus, to get at the heart of our metaproblems with life, we must adapt a critical questioning, a standing back from, a reflection that involves an object and subject of reflection, a secondary reflection. Marcel's secondary reflection involves a meta- movement, a reflection upon reflection, a reflection on how I reflect at all.

Marcel proposes that if one turns one's reflection upon one's own inner world, one finds an uneasiness about life and living, a metaphysical uneasiness, an ontological exigence, a lack of some*thing*. This lack sears through one's inner world because one ignores some appeal sounding from the depths of one's being. Marcel suggests that this already-there calling is not always heard, not always recognized, and not always acknowledged. Some people in the modern world interpret this lack as a life "signifying nothing."[136] Marcel interprets this lack as a problem of being. Because one does not understand who one is, one experiences restlessness in the world and within. Marcel suggests that if one were to confront this lack of union within, one would encounter one's own being as called upon, as ontological exigence. At the heart of experience, one feels this longing to be with others, an emptiness that calls out *for* acknowledgment. Confronting this lack of union within is an awakening to being.

Marcel's philosophy of being was his acknowledgment of having personally heard an appeal of being and having seen the tragedy of a technological world that seems to ignore this appeal. The tragedies brought on by technology require responses of tremendous proportion to fill the gaping wounds etched throughout human relations to life, to environment, and to one another. One gaping wound appears in *intra*personal worlds. These metaproblems of being on a personal level require solutions that address being on a personal level. The next two chapters engage a *meta*-technological response by turning from technological reflections to foreground Marcel's reflection on *being*. Awakening to being begins by recognizing the sound of being as *how I am* in the world. In the next chapter, this work explores Marcel's reflection upon being as being in its most concrete, grounded, enfleshed form: being as body: what does it mean *that* I am?

At the heart of our problems in being *with* others in this world, we find our problems with being at all in this world. At the heart of our interpersonal struggles to communicate, we find our intrapersonal struggles to understand who I am. Questions echoing as intrapersonal dialogues reveal personal uncertainties about communion. Healing of this internal disunion sets the stage for healing who I am as a communicator. Awakening to being, one begins a critical distancing, a metatechnological distancing to get at the heart of one's problems in being at all, an awakening to where, how, and who one is.

NOTES

1. Gabriel Marcel, "On the Ontological Mystery," in *The Philosophy of Existentialism*, translated by Manya Harari, 9–46 (New York: Citadel Press Kensington Publishing Corp, 2002), 30.
2. Ibid., 31.
3. Ibid., 30.
4. Ibid., 30.
5. Gabriel Marcel, *The Mystery of Being: Faith and Reality*, vol. II, translated by René Hague (Chicago: Henry Regnery Company, 1951), 148. The quote is "Life in this world has become more and more widely looked upon as a worthless phenomenon."
6. Gabriel Marcel, *Being and Having: An Existentialist Diary*, translated by Katherine Farrer (New York: Harper and Row, 1965), 186.

7. Friedrich Nietzsche, *The Gay Science With a Prelude in German Rhymes*, translated by Josefine Nauckhoff (Cambridge University Press, 2001). Jean Wahl, *Philosophies of Existence: Introduction to the Basic Thought of Kierkegaard, Heidegger, Jaspers, Marcel and Sartre*, translated by F.M. Lory (London: Routledge, and Kegan Paul, 1969). Sarah Bakewell, *At the Existentialist Café: Freedom, Being, and Apricot Cocktails with Jean-Paul Sartre, Simone de Beauvoir, Albert Camus, Martin Heidegger, Maurice Merleau Ponty and Others* (New York: Other Press, 2016), 20.
8. Jean Wahl, *Philosophies of Existence*, 29.
9. Marcel, "On the Ontological Mystery," 12.
10. Gabriel Marcel, *The Mystery of Being: Reflection and Mystery* vol. I, translated by G.S. Fraser (South Bend, IN: St. Augustine's Press, 2001), 22.
11. Marcel, "The Broken World," in *Gabriel Marcel's Perspectives on The Broken World*, translated and edited by Katherine Rose Hanley, 31–152 (Milwaukee: Marquette University Press, 1998), 46.
12. Marcel, *Mystery of Being*, vol. I, 162.
13. Marcel, "On the Ontological Mystery," 12.
14. Gabriel Marcel, *The Existential Background of Human Dignity* (Cambridge, MA: Harvard University Press, 1963), 160.
15. Gabriel Marcel, *Man Against Mass Society*, translated by G. S. Fraser (Chicago: Henry Regnery Company, 2008), 44.
16. Ibid., 10.
17. Ibid., 1, 151–152. Marcel notes how nations continue to relate to people through "military machine" brings a weakening of human bonds as a "growing depersonalization of human relations."
18. Vincent P. Miceli, *Ascent to Being: Gabriel Marcel's Philosophy of Communion* (New York: Desclee Company, 1965), 92–93.
19. Marcel, "On the Ontological Mystery," 26.
20. Ibid., 26–27. Marcel describes this evaluation as having a balance sheet between one's world around them and one's internal world.
21. Gabriel Marcel, *Tragic Wisdom and Beyond*, translated by Stephen Jolin and Peter McCormick, edited by John Wild (Evanston: Northwestern University Press, 1973), 12–13.
22. "On the Ontological Mystery," 26.
23. Ibid., 28.
24. Susanne K. Langer, *Philosophy in a New Key* (Cambridge, MA: Harvard University Press, 1957), 3–4.
25. Marcel, "On the Ontological Mystery," 13. Marcel states "There exists in such a world, nevertheless, an infinity of problems, since the causes are not known to us in detail and thus leave room for unlimited research. And in addition to these theoretical puzzles there are innumerable technical problems, bound up with the difficulty of knowing how the various functions, once inventoried and labelled, can be made to work together without doing one another harm. These theoretical and technical questions are interdependent, for the theoretical problems arise out of different techniques while the technical problems cannot be solved without a measure of pre-established theoretical knowledge."
26. Marcel, "On the Ontological Mystery," 16.
27. Marcel, *Being and Having*, 39. Gabriel Marcel, *Metaphysical Journal*, translated by Bernard Wall (Chicago, IL: Henry Regnery Company, 1952), 97. *Mystery of Being*, vol. I, 2.
28. Marcel, *Man Against Mass Society*, 74.
29. Marcel, *Being and Having*, 170.
30. Marcel, "On the Ontological Mystery," 16.
31. Marcel, *Being and Having*, 117.
32. Marcel, *Mystery of Being*, vol. I, 213.
33. Marcel, "On the Ontological Mystery," 178.
34. Marcel, *Creative Fidelity*, 68.

35. "On the Ontological Mystery," 180. Marcel describes a mystery as a "problem that trespasses on its own givenness: I, who inquire about the meaning and the conditions of possibility of this encounter, cannot place myself outside or opposite it."
36. Marcel, *Being and Having,* 100. Miceli, *Gabriel Marcel's Philosophy of Communion,* 96–97. Brian Treanor, *Aspects of Alterity: Levinas, Marcel, and the Contemporary Debate* (NY: Fordham University Press, 2006), 62–65.
37. Marcel, *Being and Having,* 100.
38. Ibid., 117.
39. Richard Zaner, *The Problem of Embodiment: Some Contributions to a Phenomenology of Body* (Belgium: Martinus Nijoff, 1964), 7–8. Zaner suggests that Marcel offers a "metaphysics of personal existence," a study of what it means to be, is a "metaproblem."
40. Marcel, *Being and Having,* 171. "On the Ontological Mystery," 178.
41. Marcel, "On the Ontological Mystery," 178.
42. Dwayne A. Tunstall, *Doing Philosophy Personally: Thinking about Metaphysics, Theism, and Antiblack Racism* (New York: Fordham University Press, 2013).
43. Marcel, "On the Ontological Mystery," 38.
44. Marcel's exploration of questions of human experience
45. Marcel, *Tragic Wisdom and Beyond,* 194.
46. Marcel, "On the Ontological Mystery," 31.
47. *Man Against Mass Society,* 74–75. Marcel describes this situation in 1951. This statement bears new meaning in 2020, amid the struggles by global, national, and local leaders to control the spread and devastation of the COVID-19 virus.
48. Marcel, *Tragic Wisdom and Beyond,* 194.
49. Ibid., 195
50. Ibid.
51. Gabriel Marcel, *The Decline of Wisdom,* translated by Manya Harari (New York: Philosophical Library, Inc., 1955), 13–14.
52. Marcel, *The Decline of Wisdom,* 15. This work published in 1953 does not begin to account for the catastrophes resulting from globalization and industrialization in countries without concern for people, environments, or futures.
53. Ibid.
54. At the time of my writing this, my mind turns to our current experience of the COVID-19 pandemic. A pandemic for which we should and could have planned for more effective responses. In this instance, international technological planning would have given nations and communities strategies for addressing these problems. as has been proven in vast, international failures to respond well to it—to the 2020 pandemic.
55. In the case of the COVID-19 pandemic, not only do (did) we need technological solutions, but we need to address human problems of trust in public health experts, communal responsibility for all people, and individual commitment to abide by health guidelines. I am not offering simplistic descriptions of the many complexities experienced in a global pandemic. I am not suggesting that we could with some serious reflection have avoided or been ready for these events of 2020. I am just using this as an example of a life event that quickly became a metaproblem for our world. I will not draw out any more this comparison of Marcel's description of a metaproblem to our current situation with COVID-19 in 2020 when I am writing this. I move back to our use of technologies in a more personal, interpersonal, and communal level, as spelled out in Chapter Two of this work.
56. Marcel, *Tragic Wisdom and Beyond,* 199.
57. These three areas are vastly different areas that deserve more nuanced study of what technological progress in this area means for the human person. Unfortunately, these discussions go beyond the scope of this project.
58. Marcel, *The Decline of Wisdom,* 19.
59. Ibid.
60. Lewis Mumford, "The Technological Imperative," in *The Meaning of Technology: Selected Readings from American Source,* edited by Montserrat Ginés, 129–131 (Barcelo-

na, Spain: Centre de Publicaciones del Campus Nord, 2003), 129–30. Mumford describes this technological imperative in *The Pentagon of Power*, "Western society has accepted as unquestionable a technological imperative that is quite as arbitrary as the most primitive taboo: not merely the duty to foster invention and constantly to create technological novelties, but equally the duty to surrender to these novelties unconditionally, just because they are offered, without respect to their human consequences."

61. Marcel, *Tragic Wisdom and Beyond,* 14.

62. Marcel,"Testimony and Existentialism," in *The Philosophy of Existentialism*, Gabriel Marcel, translated by Manya Harari, 91–103 (New York: Citadel Press Kensington Publishing Corp, 2002), 102. Throughout his writing, Marcel weighs in on the differences between his philosophy and Sartre's philosophy. In "Testimony and Existentialism," Marcel offers a comparison of his ideas with Sartre's approach to the study of existence. Gabriel Marcel, *Homo Viator: Introduction to a Metaphysic of Hope,* translated by Emma Crawford (New York: Harper and Row, 1962), 166–184. In *Homo Viator,* Marcel reflects at length on Sartre's ideas in *Being and Nothingness.*

63. Marcel,"Testimony and Existentialism," in *The Philosophy of Existentialism*, Gabriel Marcel, translated by Manya Harari, 91–103 (New York: Citadel Press Kensington Publishing Corp, 2002), 102.

64. Marcel, *Man Against Mass Society,* 195.

65. Marcel, *Homo Viator,* 34.

66. Ibid., 45, 46.

67. *Man Against Mass Society,* 14.

68. *Mystery of Being,* vol. I, 64.

69. Donald F. Traub, *Toward a Fraternal Society: A Study of Gabriel Marcel's Approach to Being, Technology, and Intersubjectivity* (New York: Peter Lang, 1988), 183.

70. *Mystery of Being,* vol. I, 77.

71. Ibid., 78.

72. Richard Zaner, *The Problem of Embodiment,* 6. Zaner uses the original phrases in French offers his translations as first reflection and second reflection. Other translations of Marcel use "primary reflection" and "secondary reflection".

73. Marcel, *Mystery of Being,* vol. I, xi.

74. Treanor, *Aspects of Alterity,* 65.

75. Marcel, *Mystery of Being,* vol. I, 83.

76. Traub, *Toward a Fraternal Society,* 82.

77. Marcel, *Mystery of Being,* vol. I, x.

78. Marcel, *Tragic Wisdom and Beyond,* 235.

79. Tunstall, *Doing Philosophy Personally,* 1–18. Tunstall offers an excellent summary of Marcel's approach to the study of being and how Marcel's approach can be understood alongside contemporary philosophical views of the study of being.

80. Marcel, "On the Ontological Mystery," 16.

81. Ibid., 176.

82. Ibid., 16.

83. Ibid., 6.

84. Ibid., 177.

85. Marcel, *Mystery of Being,* vol. I, 125. "The unity of the situation appears to those 'involved' in it as essentially being a datum given [to consciousness] but at the same time as something that permits of and even calls for their active intervention." *Metaphysical Journal,* 137.

86. Marcel, *Mystery of Being,* vol. I, 101.

87. Marcel, "An Essay in Autobiography," in *The Philosophy of Existentialism,* translated by Manya Harari (New York: Citadel Press Kensington Publishing Corp, 2002), 127–128. Marcel states the main thrust of his work: "perhaps I can best explain my continual and central metaphysical preoccupation by saying that my aim was to discover how a subject, in his actual capacity as subject, is related to a reality which cannot in this context be regarded as objective, yet which is persistently required and

recognized as real." And yet, the philosopher as subject, "can never stand in the relationship of an onlooker to a picture."

88. Marcel, "On the Ontological Mystery," 17.
89. Marcel, "An Essay in Autobiography," 128.
90. Marcel, "On the Ontological Mystery," 177.
91. Marcel, *Being and Having*, 171.
92. Gabriel Marcel, *Philosophical Fragments: 1909–1914*, translated by Lionel A. Blain (South Bend, IN: The University of Notre Dame Press, 1965), 84.
93. Marcel, "On the Ontological Mystery," 14.
94. Henry G. Bugbee Jr., "L'Exigence Ontologique," in *The Philosophy of Gabriel Marcel, Library of Living Philosophers Volume XVII*, edited by P.E. Schillp and L.E.Hahn (Open Court Publishing, 1984), 92.
95. Tunstall, *Doing Philosophy Personally*, 17.
96. Margaret M. Mullan, "Gabriel Marcel's Approach to Recognizing presence: Being, Body, and Invocation," *Review of Communication* vol. 18, no. 4 (2018), 319–335, 324.
97. Marcel, *Existential Background*, 79.
98. Marcel, *Being and Having*, 118.
99. Ibid.
100. Marcel, *Metaphysical Journal*, viii.
101. Ibid., x.
102. Ibid.
103. Marcel, *Existential Background* 75, 79.
104. Marcel, *Being and Having*, 118.
105. Marcel, "On the Ontological Mystery," 23.
106. Ibid., 25.
107. Ibid., 23.
108. Marcel, *Being and Having*, 118.
109. Tunstall, *Doing Philosophy Personally*, 39.
110. Ibid.
111. Marcel, "On the Ontological Mystery," 26.
112. Ibid., 25.
113. Ibid., 24.
114. Zaner, *Problem of Embodiment*, 19.
115. Marcel, "On the Ontological Mystery," 181.
116. "On the Ontological Mystery," 26. Marcel considers the act of secondary reflection as "the most vital . . . the most dramatic moment in the rhythm of consciousness seeking to be conscious of itself."
117. Marcel, *Existential Background*, 82–83.
118. Ibid., 83.
119. Marcel, *Tragic Wisdom and Beyond*, 54.
120. Marcel, *Mystery of Being*, vol. II, 40.
121. Marcel, *Homo Viator*, 137.
122. Ibid., 138.
123. Marcel, *The Decline of Wisdom*, 16.
124. Marcel, *Tragic Wisdom and Beyond*, 51.
125. Ibid., 54.
126. Marcel, *Mystery of Being*, vol. II, 37. (Marcel's emphasis)
127. Ibid.
128. Marcel, *Being and Having*, 125.
129. Marcel, "On the Ontological Mystery," 15.
130. Ibid., 27.
131. Ibid., 38.
132. Marcel, *Mystery of Being*, vol. II, 45.
133. Ibid., 46.
134. Marcel, *Tragic Wisdom and Beyond*, 51–52.

135. I offer this qualifier to note that I do not believe comprehensive self-understanding is possible. There may be persons who are exceptions to this experience and who do fully understand themselves. I just note that I have not met this type of person, yet.

136. Marcel does not use these exact words, he describes nihilism as both signal and cause of this despair. This work cites William Shakespeare's words capturing this type of nihilism, "Life's but a walking shadow; a poor player, that struts and frets his hour upon the stage, and then is heard no more: it is a tale told by an idiot, full of sound and fury, signifying nothing" (Shakespeare, *Macbeth,* v. 17–28, p. 540).

FOUR

Positioning for Communication

Gabriel Marcel's Philosophy of Body

Before we may attempt to communicate with those near us, we become aware that we are near someone else, who is able to communicate with us. Asking "can you hear me now?," we ascertain if the other person is actually here and now, able to open channels for communication. A response, yes, I am here, I can see, hear, or touch you lets us know that we can communicate. The dimensions of nearness or hereness constitute crucial ground for possible communication. Sensing *that* I am here and *that* you are here assumes vital importance for communicators. Becoming aware *that* I or you are here means becoming aware of my/your body. When we overlook body or bodies, we lose awareness *that* we are here and now and we have no ground for beginning communication. Gabriel Marcel invites reflection on one's experience: I am here. Building upon our search to understand "who I am" (as discussed in Chapter Three), in Chapter Four, we seek to understand *how* I am. Marcel's call for reflection on body as *how* I am uncovers the vital role of positioning *for* communication.

Some of the problems in modern living are rooted in overlooking the importance of body. Privileging one's mind and ideas over one's feelings or vital experience becomes a problem of ignoring body as a valid source of experience. In *The Problem of Embodiment,* Richard Zaner suggested that Gabriel Marcel, Jean-Paul Sartre, and Maurice Merleau-Ponty each offer important theories about the human body. Marcel builds his philosophical reflections around our most foundational experience as body.[1] For Marcel, René Descartes' privileging *cogito* as the sole way one knows the world is an "oversimplification" of how one knows one's world.[2] The *sum* (I am) in Descartes' "cogito ergo sum" does not take into account the

"I am" who seeks to understand her own existence.[3] As Chapter Two elucidates, focusing solely on mental activities may lead to abstractions about the human person that dismiss key bodily dimensions of experience. These abstractions of the human person to her mental output or productivity can lead to mistreatment—sometimes tragic mistreatment—of those human persons who do not produce enough in intellectual realms. This idealist understanding of human experience negates the importance of looking to existence to describe our relations in our world.[4] Some versions of this idealist study of life become solipsistic and ineffective (and sometimes harmful) in helping one understand the reality of one's own existence.[5]

Marcel calls for a study of both our mental and vital experience.[6] Marcel finds an intimate connection between thinking about life and living one's life. The art of reflection is "one of the ways in which life manifests itself."[7] One participates in living in a deeper way when one reflects on the meaning of that life. But, Marcel qualifies, this reflection on life must not be limited to reflecting on one's mind. Reflection attends to living experience: an experience happening as mental and bodily experience. One should not study life as a general experience but one should study life lived by "a particular individual."[8] Marcel calls for grounding or beginning one's study of the world in the "touchstone of existence," one's "Here I am," one's "I exist."[9] Vital experience is lived experience, felt, sensed, lived as body. In 1914, Marcel's suggestion of body-as-mine as a touchstone of existence precedes the discussions about embodiment offered by other early twentieth-century philosophers.[10] To counter problems of abstractions about human beings, Marcel calls for focusing on concrete experience of being human, which appears in its most concrete form as body.

Life, as bodily experience, communicates meaning. The problem in the modern world is that many people do not reflect on, nor know what "my body" means. Marcel's philosophy of body explores what one's relationship with body is and means. This chapter attends to the question, what does the concrete human experience of "being as body" mean? Marcel offers the term incarnate being to describe life lived as body, as this, and as my body. First, this chapter explores Marcel's description of the feeling and meaning of incarnate being. Second, this chapter extracts from Marcel's philosophy of body *how* we experience incarnate being through relations of being and having, as participation, and as ground for freedom. "Being as body" is *how* I am in the world. "Being as body" means being in relation to. Being in relation to indicates positioning toward or away from the others who are also in this world. "Being as body" reveals possibilities for positioning toward and participation with as *how* I am in the world. This chapter explores what Marcel's "being as body" might mean for communication.

"HERE I AM!" REVEALS *THAT* I AM INCARNATE BEING

Marcel's philosophy, as exploration of what life is, starts in his reflexive questioning, "Who am I—I who question being?"[11] He first asks, "This I, I who ask questions about being, can I be sure that I exist?"[12] As noted above, he rejects Cartesian separation of mind from life and seeks to maintain a "unity of a living subject" who thinks and thinks about herself.[13] How does one think about how I exist, how I am in the world? Marcel states, "the whole reflexive process remains within a certain assertion which I *am*-—rather than *which I pronounce*—an assertion of which I am the place, and not the subject."[14] The very act of reflection can be enacted because one *is*, "I am the place." Existence precedes reflection: one *is* before one *thinks*, before one *thinks about* how one is. Marcel's philosophy suggests that one understands who one is through awakening to one's experience as incarnate being.

Marcel seeks to understand "Who am I" not so much that I am. Zaner describes Marcel's search as a "quest for assuredness about myself."[15] What does it mean that I am *who* I am—I who question being. This "Who am I—I who question being?"" forms a metaproblem: a question that also requires attending to the question-er. To seek responses to the question, "Who am I?" one engages a method of knowing phenomena in the world that is more than rational thought. Marcel suggests one can assert, "I exist" "only if it signifies, in an admittedly loose and inadequate way, an original datum which is not 'I think' nor even 'I am alive,' but rather 'I experience.'"[16] Marcel suggests that in the action "I experience" there is more than a subject who thinks about or who possesses her own life. As one reflects on the fact that "I am," one uncovers that one is a being who has feelings.[17] The questioner feels[18] *that* she exists before she understands that existence. "I experience" involves body, feelings, and making sense of (or understanding) these phenomena. One becomes aware of these "facts of existence" through vital feelings and through reflection about this data.[19]

One's entry into existence is being born to someone. Incarnate being from the beginning of existence is in relation to other. Marcel connects birth into existence to body: "I am the incarnate reply to the reciprocal appeal between two beings."[20] An incarnate being emerges in response to presence of another. One is born into community and comes into existence in the presence of others.[21] Incarnate being begins by being with.

Incarnate being is how one is manifest in the world and so also bears dimensions as body, as extension in space. Marcel describes a "Here I am" that communicates "hereness and nowness" of this person.[22] Marcel connects one's corporeity to one's history. Corporeity "implies what we may call historicity. A body is a history, or more accurately, it is the outcome, the fixation of a history."[23] My body is a history. Marcel does not expand upon the meaning of this phrase in this particular discussion.[24] Body communicates that one *was* in past times and that one *is* in the world. Marcel does discuss how one is not only thrust into a historical situation but is also connected in some way to those who go before one.

Marcel explores, what does it mean that I am who I am. One's sense of body is intimately or intrinsically tied to one's awareness of existence. One's *that* I am is united to—and *is*—who I am. Body is one's "loci and moment," one's "hereness and nowness," one's "ecceity" or one's "immediate experience."[25] Since his 1914 writings in *Metaphysical Journal,* Marcel explores how body relates to existence and consciousness. Consciousness exists "only in the measure in which it is given in an immediate relation to itself or to another."[26] I am conscious of "*my body.*"[27] Consciousness exists in relation to *my body.*

> When I assert: I exist, I certainly mean something more than [pure feeling]; I vaguely imply that I am not only for myself but that I manifest myself, or rather am manifested; the prefix *ex* in exist, has primary significance because it conveys the meaning of a movement towards the external world, a centrifugal tendency. I exist: that means I have something by which I can be known or identified either by another person or by myself insofar as I assume for myself a borrowed otherness; none of these characteristics are separable from the fact that 'there is my body.'[28]

One perceives one's body as being in physical space, one is conscious of one's physical arms and legs. In the cry "Here I am," I am "manifest," I exist—with the word exist including the "Latin prefix *ex*—meaning *out, outwards, out from.*"[29] Body, this body, in extension in space, manifests or communicates existence. I am manifest or I am manifested signals existence in relation to other; others may know and identify me because I am manifest, existing as my body. Marcel describes this as "borrowed otherness," as a being other in this being manifest to. Much like, incarnate being is a reply to, in "borrowed otherness," I exist as other to others. Marcel's interpretation of existence includes an awareness that one is in the world as body, as embodied, and as manifest to other.

If body is a touchstone of existence, it is only ever a foundational measurement experienced first-hand, as my body.[30] One's situation in the world is always already as body. One's material flesh and living organism located here and now *is*. Marcel named this situation incarnate being: "our very mode of existence as incarnate beings."[31] This state of "incarnation[32] is the situation of a being who appears to himself to be, as it were, *bound* to a body."[33] Incarnate being means that one always is as this "particular body" as "my body."[34] One exists always already as incarnate being. Marcel, attempting to put into words how one relates to one's body, calls the ground of our existence "Incarnation, the central given of metaphysic."[35] Incarnate being is how one can speak about existence at all. As ground for existence, incarnate being is how one is fundamentally situated in the world.[36] Garth Gillian describes Marcel's phrase incarnate being as Marcel's category of facticity, what Martin Heidegger would call "thrownness."[37] Marcel identifies incarnate being as ground, as touchstone and landmark of existence.[38]

Body not only communicates *that* one is here and now; body is *how* one is here and now. Marcel suggests that body is in fact, a "touchstone of existence."[39] One's body constitutes "datum on which everything else hinges."[40] All sensation or knowledge of information about anything in the world appears as or through data that is one's body. We have knowledge of our world because our body constitutes a "touchstone of existence." One has a sense of being in the world through one's relating to all in that world. All relating is possible because "I am *incarnate*."[41] I acknowledge existence because I-as-body can: "when I affirm that something exists, I always mean that I consider that something as connected with my body, as able to be put in contact with it, however indirect this contact may be."[42] All relating is possible because I am incarnate being.

Even one's way of relating to self[43] is possible because one is incarnate being. One relates to oneself on a fundamentally intimate level as how one feels about oneself. Marcel identifies affective dimensions in the realization that one is incarnate being. One has an "*Urgefühl* or primordial feeling,"[44] that one's body is "intimately mine."[45] Marcel suggests that these feelings are felt pre-reflectively or as an "*a priori* of pure sensibility."[46] One's sense of incarnate being is a primordial feeling or foundational way of relating in and to the world.[47] Much like intuition of being described in Chapter Three, one realizes—through secondary reflection—*that* one is incarnate being, bound to body and in relation to world as body.

One experiences incarnate being on a vital level, as a feeling of one body as my body. Feelings are connected to being incarnate being: "I am my body only in so far as I am a being that has feelings."[48] Marcel's use of the word "sentir" points back to his earlier explanation in *Metaphysical Journal*, "between consciousness and body there is another relation inasmuch as my body is a datum given to internal perception," in which one "coenesthetically" perceives one's bodily sensations.[49] Feelings of coenesthesia include experiences like "being tired, hungry, energetic" in which one feels tiredness in and as my body.[50] Coenesthetic feelings of internal perception about one's body could be called body-consciousness or "ontological perception."[51] One feels body as mine if one is able to feel "coenesthetically." One experiences incarnate being on a vital level, on a level of feelings.

Marcel attempts to describe pre-reflective feeling about oneself using the phrase "sympathetic mediation" as meaning that one feels some feeling or experience without a conscious acknowledgment of one's feeling.[52] This "sympathetic mediation" indicates our "non-instrumental communion with our bodies."[53] Marcel noted "sympathetic mediation" was the least inadequate description for this relation. Zaner finds Marcel's phrase "sympathetic mediation" problematic. Instead of using this phrase to explain how *Urgefühl* "mediates" my body to me, Zaner suggests saying my body "embodies me."[54] Zaner argues, saying "I experience this body as mine, is to say that it embodies me immediately."[55] Embodiment itself is a foundational relationship, not a relation of mediation.

I experience my body as mine on this vital level, at the heart of how I relate at all to anything or anyone in the world. Marcel suggests that one awakens to this reality of incarnate being or body as mine through realizing one's foundational, pre-reflective way of being. Thus, Zaner concludes that Marcel's term incarnate being means a "mystery of *embodiment*, my *être-incarnée*."[56] One's relation to the mystery of being incarnate is an interaction that is more than a subject knowing an object. To understand one's own relation as being incarnate being one must move away from a subject-object relation.

To understand one's own relation to body, one must understand how being itself involves participation which is both active and passive/receptive. Marcel attempts to distinguish this relation to body as body-object and body-subject. Incarnate being involves:

> This given is opaque to itself, in opposition to the *cogito*. Of this body, I can neither say that it is I, nor that it is not I, nor that it is *for* me (like an object). The opposition of subject and object is found to be transcended from the start.[57]

As a given that is not always clearly decipherable, body also resists being defined as objective types of relations. Marcel notes that one cannot say one *is* her body nor that one possesses her body; thus, body is neither fully subject nor all object.[58]

Marcel, warning against assuming a subject-to-object relation to one's body, suggests a body-object and body-subject distinction. Although my body as mine belongs to me, this belonging is not a total: "my body belongs and does not belong to me."[59] We cannot convert body into an object.[60] One relates to one's own body in some ways as being other than oneself. If another person looks at my body and views it as an object, I am not reduced to that objectified body viewed by the other. Because one's body is not fully identified with "me," one's body is sometimes experienced as an object: as "body-object."[61] One's body that is seen by others as flesh in dimension is in a sense body-object or that which is seen. And if one loses a limb, one's relation to the dismembered body part is one of body-object. One engages one's body as subject when one expresses that "my body is mine, I do not place any interval between it and me, or rather for as much as it is not an *object* for me, but in so far as I *am* my body."[62] The "I" who experiences "my body" is body-subject. Awakening to my body-as-mine is an experience of body-subject. Marcel signals his view of embodiment when he describes "the situation of a being who appears to himself as fundamentally and not accidentally connected to *his* body."[63] One is intimately and existentially connected to his body as his. In embodiment, I become aware of my body as mine.

One feels that one is incarnate being as an act of participation. One who feels this deeper meaning as body does not mean that body as sensing organism simply receives and transmits external stimuli.[64] When one recognizes, through secondary reflection, that one's body is mine, one feels that this body is mine in an unmediated, immediate way. One feels that one is incarnate being as an act of "communion" or involved participation.[65] This vital experience of being in communion or participating *with* reveals meaning at the heart of one's life. Being incarnate being in the world springs from or dwells within one's most intimate, inner world. Awakening to being incarnate being is an awakening to one's core as being in the world.

HOW I PARTICIPATE IN THE WORLD: AS INCARNATE BEING

The body offers a point of contact for all relations, "between me and all that exists there is a relation (the word is quite inadequate) of the same type that unites me to my body."[66] Marcel posits one's relation to one's world mirrors one's relation to one's own body. As body, one knows any and all beings in the world. Marcel sought categories to describe how one exists in the world. He proposed that body offers insights into these categories of existence. As noted, this existence as incarnate being is a participation in being in the world. One participates in the world as incarnate being which means *as* my body as mine and *as* my body as presence. These dimensions of incarnate being reveal how one participates in the world through both being and having. Marcel's philosophy of body explores *how* we experience incarnate being through relations of being and having, as participation, and as ground for freedom.

One is born into existence, as incarnate being, as such, one is always already in relation to other and body is always already experienced by me as mine. Sensing that my body as mine may be a *Urgefül*, a primordial feeling, but one requires secondary reflection to understand what my body as mine reveals to me about being in the world. Reflection upon relation of body as mine uncovers a relation of *being and having* body. Throughout his work, and particularly in his 1935 *Being and Having: An Existentialist Diary,* Marcel explores being and having as significant ways people relate to phenomena in their world. In *Being and Having,* Marcel suggested, my body is a prototype for how I have.[67] As prototype for having, the way one *has* one's body informs how one *has* in general. Marcel's description of the verbs being and having need not be set against one another but may be engaged along a continuum. James Collins found that Marcel's description of relating to human reality included a "unique complexus of having and being."[68] Collins explained this "complexus" as allowing one to exist independently of being reduced to an instrument or possession while also possessing a materiality that could not be abstracted to "pure being."[69] Having and being one's body emerge along a spectrum:

> My body is *my* body just in so far as I do *not* consider it in this detached fashion, do not put a gap between myself and it. To put this point in another way, my body is mine in so far as for me my body is not an object, but rather, I *am* my body.[70]

Marcel's statement "I *am* my body" presented some confusion about what Marcel meant by this relation of being and body. Zaner addresses this ambiguity in Marcel's statement by exploring why Marcel might have said this. One always experiences body as mine so while reflecting on body as mine "it presents itself as *already mine.*" Zaner argues that it is "misleading, to say that 'I *am* my body'" because this confuses the two ways one understands that my body is mine. One way I mean that my body is mine is an automatic coenesthetic sense that this living organism is mine. A second way I mean that my body is mine is through an affirmation about incarnate being. Zaner suggests that these are clearer ways of describing what my body-as-mine means. Marcel's exploration of the meaning of my body as mine sheds light on the meaning of incarnate being.

Body as mine indicates that one in some sense possesses her body. Marcel used the verb "belonging to" or body as mine to depict this relation.[71] We consider our body to belong to us. This sense of body belonging to me indicates a place within which one dwells.[72] Having my body as mine implies "being able to dispose of, having power over."[73] Marcel saw body is a "unified context of powers," as in each power that I have of moving muscles or embracing someone is an "assemblage of powers" rooted in my having my body as mine.[74] Because the body manifests these powers, the body is one's site for acting in and upon the world. One's ability to use this body as *my* body reveals one's power over the body, a power "to extend the body itself."[75] Thus, one engages one's power over one's body, is manifesting oneself through kinesthetic movements, as incarnate being *in* extension, in hereness and nowness in the world. To be in the world means to feel and have power as one's body as mine.

Marcel warned against reducing incarnate being to only *having* a body. Having my body as mine does not mean one may treat one's body as one's instrument. In his early work, Marcel described how one might feel that one could relate to one's body as an instrument to be used.[76] One might attempt to treat one's body as distant from I who am, to create a distance—a mentally constructed abstraction—between "I" and my body. Marcel suggests that this attempted view of body-as-instrument engages primary reflection. Using primary reflection to study body reduces the body to an object that I am using as instrument. Yet one cannot relate to one's body as object because "my body is posited as the center in relation to which my experience and my universe are ordered."[77] As "center in relation," body positions one in relation to all other entities in the world. Thus, one cannot ever create distance from the body to treat the body as an object. While one may view the body as a constituted by parts such as bones, limbs, or organs, one cannot create distance from the body felt as a unity, my body. Marcel calls the idea that one could use body as one's instrument a "pseudo-idea."[78] My body-as-mine is something I possess and yet cannot be one's instrument.[79] Because one is incarnate being, there exists a "community of nature" as one is always already as one's body and one always already has body as mine.[80] Yes, incarnate being indicates body as mine, but this possession of body is not a possession that permits instrumentalizing body.

Although one has the power to act in one's body, Marcel qualified that one does not wield full possession of one's body. One might *try* to relate to all phenomena in the world as things to be possessed. Yet if one attempts to relate to one's own body as some *thing* to possess, something happens in the relation. When I treats my body as if it were my possession, the body exercises a "tyranny" over me.[81] One's body resists possession. One "cannot say that [my body] is at my disposal" because one cannot dispose of "that which gives me the disposal of things."[82] The owner of the body-as-given does not have complete power over the body but, in some way, receives the body. The body that gives me power to dispose of things cannot be fully possessed. Marcel offered the example of suicide to depict when I experience limits in relating to my body as at my disposal. Every person has the power to dispose of one's body through killing oneself. This power to dispose then puts one in a state of being completely dispossessed of one's body and thus powerless to continue disposing of one's own body. Use of one's power to dispose of one's body leads to loss of all power over one's body. In suicide, one thinks they are acting in full power over body. To Marcel, suicide communicates the limits of acting as if one has full power to do what one wants to one's body. Body "encroaches" upon me.[83] Dimensions of one's body lie outside a realm of possession, and the body-as-given acts upon or is present to the embodied one.

One who *is before* something is not in full control of that something. We experience body as revealing realms overabounding one's sense of possession of body as only mine. As incarnate being I am in the presence of body as other, I am *with*. In reflecting on one's own body, one comes into a presence. Much like Marcel's explanation of the act of recollection, this reflection on how one is as incarnate being: one abandons oneself, comes into the presence of being incarnate being as a phenomenon to be acknowledged not possessed. One experiences being "in-the-presence-of" my body as "something other than itself."[84] Marcel called this realm that could not be possessed "a presence."[85] He was not conflating a presence with an object. The presence of body is that which is recognized by an other as being other.[86] Marcel's works suggest that "my body is apprehended as a presence; the thou is given to me as co-presence."[87] Experience of my body, being-as-presence, involves being-before that which is other.

While body resists a relation of possession, incarnate being invites a distinct relation between me and my body. As noted earlier in this chapter, one feels that one is incarnate being as an act of "communion" or involved participation.[88] Participation indicates a relation of being with or being receptive to. Marcel uses the term incarnate being for how one participates in world, how one *is* in the world.[89] I participate in the world as "my presence" in this world, as "nexus of my presence to the world, my body being this nexus manifested."[90] Marcel connects participation, being, and presence. Marcel connects incarnate being with a receptive way of being in the world. If one becomes aware of how one participates in the world, one opens oneself to another to welcome the other as other.[91] As incarnate being, one is participating fully as being in the world.

Experiencing my body as mine becomes ground for freedom. Having my body as my body may imply "*to have for one's self,* to keep for one's self, to hide."[92] One relates to one's body as an internal intimate realm that one can keep to oneself. Marcel compares this relation of "keeping for one's self" is like having a secret.[93] When one *has* a secret, in one's intrapersonal realm one holds or withholds thoughts, expressions, or memories. Each person has a realm that feels internal to one's body, as if it were a realm in space within which one holds or withholds communication. Having a body gives me a sense of internal/external realms, a spatial sense of an "opposition of within and without."[94] One is free to keep one's thoughts or feelings or world to oneself or not. This capacity to withhold or not reveals body as *how* I am free.

However, incarnate being also manifests—consciously or unconsciously—to the world "Here I am!" One is manifest as my body—existing, here, and now in extension. If in one sense, I feel free to have my body as mine for myself or not, in another sense, I am not fully in control of my incarnate being. As my body, as my "Here I am!" I am manifested in the world without my agreeing to it. Marcel describes this dimension of incarnate being as "to-be-exposable."[95] As being in the world, I am manifested to others, exposed to others as body, my body. I am made present as incarnate being in this setting, this moment, and this place. "I exist: that means I have something by which I can be known or identified . . . a borrowed otherness."[96] Incarnate being means manifestation in the world that cannot be silenced (unless killed). Every incarnate being born into our world proclaims, "Here I am!" from their moment of entry into existence and during every moment since. Is this proclamation heard?

Marcel contends that in our modern world these proclamations from and of incarnate being as living presence are mostly ignored. Feeling loss of control about one's own life, one's own attitude toward life "tends to shrink back on itself itself."[97] One is not aware of one's own intimate bond to existence nor what it means to be in the world. According to Marcel, every person assumes a "position" toward how one is in the world, a position that forms the basis of our being free to act in our world.[98] A position, pointing at an embodied, spatially-located orientation, also denotes an attitude toward life, toward being in the world. One's attitude toward oneself and others can be open to or closed off from. In the broken world many people have adopted a view of my life as a closing off from. Marcel promotes refusing to accept a closed interpretation of my life—as a shrinking back on itself, closed in and withering. This refusal to accept this interpretation impels one to seek an *other* interpretation of life, to see life is more than insecurities, fears, or lack of control. If one opens oneself to life as being more than material, bodily dimensions, one opens oneself to incarnate being.

Marcel proposes secondary reflection to uncover an *other* meaning, incarnate being, of one's own life. Through attending to vital experience, one awakens to incarnate being that reveals what one's life means. Aware of incarnate being, one understands that my body is mine but cannot be disposed of, instrumentalized, or ignored without consequences. Aware of incarnate being, one is in the world through participation and may choose a position toward one's life in that world. If one positions oneself in openness toward one's own life, one also becomes more open to the life of the other.

SUMMARY: POSITIONING FOR COMMUNION

At the root of understanding oneself intrapersonally, Marcel's question "Who am I—I who question being?" points to realizing *that* I am and to understanding *how* I am. Incarnate being as *"being in the world"* is *how* I am. Realization *that* I am incarnate being reveals that one is always already in position to others. As incarnate being one is able to participate *in* this world and is already in position to others in this world. *That* I am as incarnate being opens up possibilities for communication.

First, as body, one is able to participate in communication. Marcel describes incarnate being as participation "not as a relation or communication."[99] This project proposes an amendment to Marcel's descriptor participation to include a relation of communication. Marcel's use of the word participation highlights the impossibility of subject-object relation between an incarnate being and her world. His suggestion that an act of participation is *not* communication reflects a dated view of communication from the 1940s. Several times in his works, Marcel described communication as mere transmission.[100] Scholars today, seventy years later than 1940, describe communication as much more than transmission. Communication is a process of sharing of meaning and as our ways of relating to and in our worlds—communal, interpersonal, and intrapersonal worlds.[101] Communication includes "embodied discourse" in which conscious, intentional interchanges about information are also *felt* experiences.[102] Today's descriptions amplify the meaning of communication. Communication is more than verbalized or delivered messages, communication involves more participant involvement. With this contemporary interpretation of communication, Marcel's description of participation may be opened to include communication. Incarnate being, participation, being in the world reveals a co-being with a shared mutuality of being-with. Marcel's term incarnate being contributes to our understanding of communication that may mean more.

Second, as body, one is already in position to others. One may select if one is a being who positions oneself to be open to interacting with the other, to sharing what one bears in their internal world. Even if one is physically sharing the same space and time with the other person, one may withhold one's inner world from the other. Being as body, or incarnate being, reveals that one is (partially) free to choose one's positioning toward the other. One may withhold one's thoughts or one's explanation of what one means when one acts. Being as body or in position to also points to some lack of freedom in this positioning. As incarnate beings, we always already are in communication, sounding similar to "one cannot not communicate."[103] Marcel describes this lacking freedom in position as "to be exposable to." As incarnate being, one has some freedom to select one's positioning toward or away from the other person. Marcel also describes one's freedom for positioning as attitudes toward existence.

Third, attitudes toward one's own existence shape one's attitude toward other's existence. Marcel suggests that realizing *that* one is here, "Here I am," is a wonder-filled moment. One's realization *that* one exists happens as an exclamation, as "a small child who comes up to us with shining eyes and who seems to be saying: 'Here I am! What luck.'"[104] This realization of the fact that one exists is more than a thought process arriving at a logical conclusion. This awakening is an "exclamatory awareness of existence."[105] Marcel proposes that each person bears profound feelings about one's own existence. This attitude toward life, toward my life, is similar to an existential feeling one has when one sees a newborn child or narrowly escapes death: *this* life takes on heightened meaning.[106] Awakening to this reality *that* one exists changes how one views and values one's own life. One's appreciation for life heightens one's sense that one is living here and now. As one who is always already in communication with those persons sharing one's here and now, a stronger sense of one's here and now makes one more aware of who else shares this here and now. Awareness of a shared here and now sets the ground for communication

However, even if one does not assume this wonder-filled approach to being here in the world, one always already assumes the position of incarnate being in the world. One is not in control of being acknowledged as other. As Marcel identifies, existence implies my "borrowed otherness" my incarnate being *for* or *to* others.[107] Even if one chooses to close off to others, she remains exposable as body to others. Incarnate being as the sound of presence cannot be muted. Marcel invited increased reflection on what it means to be in the world. But this call to reflection is always a response to the sound of an invocation. However deep one explores one's own intrapersonal realms, how I am in the world as incarnate being always means being *with*. Becoming aware *that* I am incarnate being grounds me in this actual world *that* is, is always already, and will be the *here and now* in which we meet the other for communication. The sound of "Here I am" as an "I am the incarnate reply to" reveals the blending of incarnate being and being *with* as participating in an ongoing song of life. Awakening to *my* being in this world allows me to hear the invocations made by my own incarnate being as the first strands of a symphony of being, a symphony which is always a being *with*, and a symphony which has already been sounding long before I became aware of my own participation.

NOTES

1. Richard Zaner, *The Problem of Embodiment: Some Contributions to a Phenomenology of Body* (Belgium: Martinus Nijoff, 1964), 3. Zaner notes "the problem of the body (or, as we shall have to say later, the *metaproblem* of the body) forms the matrix of Gabriel Marcel's philosophical work."
2. Gabriel Marcel, *The Mystery of Being: Reflection and Mystery,* vol. I, translated by G.S. Fraser (South Bend, IN: St. Augustine's Press, 2001), 87.
3. Zaner, *Problem of Embodiment,* 8–9. Zaner quotes Pietro Prini's discussion of Marcel's issues with a Cartesian starting point as an abstract *cogito.*
4. Gabriel Marcel, *Metaphysical Journal,* translated by Bernard Wall (Chicago, IL: Henry Regnery Company, 1952), 319.
5. Marcel, *Mystery of Being,* vol. I, 106.
6. Gabriel Marcel, "On the Ontological Mystery," in *The Philosophy of Existentialism,* translated by Manya Harari, 9–46 (New York: Citadel Press Kensington Publishing Corp, 2002) 17.
7. Ibid.
8. Marcel, *Mystery of Being,* vol. I, 86.
9. Marcel, *Mystery of Being,* vol. I, 90.
10. Zaner, *Problem of Embodiment,* 12.
11. Marcel, "On the Ontological Mystery," 16.
12. Gabriel Marcel, *Being and Having: An Existentialist Diary,* translated by Katherine Farrer (New York: Harper and Row, 1965), 170.
13. Ibid., 170–171.

14. Ibid., 171.
15. Zaner, *Problem of Embodiment*, 8.
16. Marcel, *Creative Fidelity*, 16.
17. Marcel, *Mystery of Being*, vol. I, 101.
18. Zaner, *Problem of Embodiment*, 35. Marcel used the word *sentir* to describe how one feels as body and how one senses things through one's body. When describing feeling, Marcel does not use the French words "*toucher*" (to touch) nor "*tâter*" (to feel). He qualified different ways that one feels. He uses the word *sentir* to describe how one experience's one's body-as-mine. As times he used the word *sentir* interchangeably with sensuous perception. *Mystery of Being*, vol. I, 104. Marcel sometimes interchanged "sensations" and feeling as he discusses one's perception and affective attitude toward the world. Zaner notes that Marcel's reflections about the body do not include a thorough explanation of sensuous perception. Marcel focused on how one feels in and through one's body. This project will use the word *feeling* for Marcel's term *sentir*.
19. Marcel, *Metaphysical Journal*, 40.
20. Gabriel Marcel, *Homo Viator: Introduction to a Metaphysic of Hope*, translated by Emma Crawford (New York: Harper & Row, 1962), 70.
21. Marcel, *Mystery of Being*, vol. I, 197–204. Marcel explores how persons interact as families or in communities of familial relation. Marcel identifies how participants in these familial communities may or may *not* assume the roles of provider(s), caretaker(s), or parent(s). This author suggests that incarnate being, as a combination of two separate person's DNA, is a heretofore unspoken response, as a unique person bearing DNA that reflects as reply to the blending of the donors DNA.
22. Marcel, *Mystery of Being*, vol. I, 175.
23. Marcel, *Being and Having*, 84.
24. Marcel's concept of my body as history hints at body being in space and time, being as extension in space and time, the facticity of incarnate being. This sounds alongside Heidegger's description of facticity as one's being bound or connected in an historical setting. Martin Heidegger, *Being and Time*, 7th ed., translated by John Macquarrie and Edward Robinson (New York: Harper Collins, 1962).
25. Marcel, *Mystery of Being*, vol. I, 109, 175.
26. Marcel, *Metaphysical Journal*, 18.
27. Ibid.
28. Marcel, *Creative Fidelity*, 17.
29. Marcel, *Mystery of Being*, vol. I, 91.
30. Oxford English Dictionary, "first hand, n., adj., and adv," *OED Online*, Oxford University Press. First hand is this authors phrase, meaning without intermediary, or in one's own person, directly.
31. Marcel, *Mystery of Being*, vol. I, 40.
32. Marcel, *Mystery of Being*, vol. I, 40. Translators used different English phrases for Marcel's French phrase "être-incarnée." G.S. Fraser translated the phrase to "incarnate being." Katherine Farrar in *Being and Having* used the word "incarnation" in translating Marcel's description of an experience. Edwin W. Straus and Michael Machado, "Marcel's Notion of Incarnate Being," in *The Philosophy of Gabriel Marcel*, vol.17., edited by P.A. Schilpp and L.E. Hahn (Chicago: Open Court, 1984). Straus and Machado use the term "incarnation." When describing the person, I will use "incarnate being" and if describing the situation, I will use "incarnation."
33. Marcel, *Being and Having*, 11.
34. Marcel, *Creative Fidelity* 20. *Mystery of Being*, vol. I, 93.
35. Marcel, *Being and Having*, 11.
36. Marcel, *Creative Fidelity*, 65. Marcel described given as a "fundamental situation which cannot strictly speaking be disposed of, surmounted or analyzed."
37. Garth J. Gillian, "Embodiment in Marcel and Merleau-Ponty," in *The Philosophy of Gabriel Marcel*, edited by Paul Arthur Schilpp and Lewis Edwin Hahn (LaSalle, IN: Open Court, 1984), 501. Gillian states that Marcel's concept incarnate being reveals "the facticity of incarnation—its manner of being a fundamental predicament—eludes

inscription into thought." Gillian continues by, "The fact of existence is obscure, and this prevents the total achievement of self-consciousness and brings the eternal world into its own ambit: 'The obscurity of the external world is a function of my own obscurity to myself; the world has no intrinsic obscurity.'" Gillian includes a quotation from Marcel's *Being and Having* that supports facticity of incarnation. I make the connection between Gillian's description and Martin Heidegger's description of "thrownness" "brought into its *there* not of its own accord." Martin Heidegger, *Being and Time*, 262.

38. Marcel, *Mystery of Being*, vol. I, 88. G.S. Fraser describes incarnate being as a "touchstone of existence." Richard Zaner uses Marcel's French word, *repère*, which I translate as landmark.

39. Ibid., 88.

40. Ibid., 91–92.

41. Marcel, *Metaphysical Journal*, 269.

42. Marcel, *Being and Having*, 10.

43. See *Homo Viator* where Marcel's discusses self, body, consciousness, and ego in his chapter "The Ego and its Relation to Others." Marcel's treatment of the term "self" goes beyond the scope of this project. Thomas Anderson, "The Nature of the Human Self According to Gabriel Marcel," *Philosophy Today*, vol. 29, no. 4 (1985), 273–283. Anderson offers clear analysis of Marcel's ideas about self. As well, see Calvin O. Schrag's *The Self after Postmodernity*, New Haven: Yale University Press, 1997, 4–9. Schrag noted complexities in modern and postmodern studies of self that incorporate the terms subject, agent, ego, identity, mind, psyche, and myself. Schrag suggested that while classical philosophers described the self-as-substance, modern philosophers viewed "self as transparent mind." Schrag opens up the meaning of self from the limited signaling of one's mind to a broader revelation of one's sense of self. Self is "who" is at the heart of this experience of one's self. Marcel's description of self sounds alongside Schrag's proposal of self as a who is at the heart of one's experience.

44. Marcel, *Metaphysical Journal*, 19, 247. Marcel differentiates Urgefühl from another internal perception coenesthetic feelings, felt in my body. Richard Zaner, *The Problem of Embodiment*, 35. Zaner suggests that Marcel's concept of "*coenesthésique*" connects to how one feels *that* one has one's body.

45. Marcel, *Metaphysical Journal*, 247.

46. Marcel, *Metaphysical Journal*, 247ff.

47. Zaner, *Problem of Embodiment*, 48. Zaner offers insightful clarification on how Marcel's writing on mediation-body could confuse the reader. Zaner notes how Marcel correctly identifies incarnate being as a "fundamental phenomenon" but sometimes incorrectly suggested that "feeling and acting mediate this embodiment." In response to Marcel's phrasing, Zaner clarifies that embodiment is not a mediated relation but one experiences one's body as mine directly or "immediately."

48. Marcel, *Mystery of Being*, vol. I, 101.

49. Marcel, *Metaphysical Journal*, 19.

50. Straus and Machado, "Marcel's Incarnate Being," 130.

51. David Appelbaum, "Body-Consciousness: Gabriel Marcel's Debt to Maine de Biran," *Bulletin de la Société Américaine de Philosophie de Langue Française*, vol. 5, no. 1 (1993), 47.

52. Marcel, *Metaphysical Journal*, 246.

53. Marcel, *Mystery of Being*, vol. I, 101.

54. Zaner, *Problem of Embodiment*, 46–48. Zaner offers critical feedback on Marcel's term "sympathetic mediation." Marcel attempting to answer the question what does it mean to call my body mine, notes how body manifests my actions in an immediate way, in "sympathetic mediation." Marcel describes feeling is a function of action and suggests an unreal "inseparability" between feeling and action. Zaner suggests that feeling is actually related to action in "*mutual foundedness*" both are functions of the other. Zaner questions if Urgefühl "mediating"my body to me is the correct description for one's relation to feeling my body as mine. Marcel noted "sympathetic media-

tion" was the least inadequate description for this relation. Zaner suggests saying my body "embodies me."

55. Zaner, *Problem of Embodiment*, 63.
56. Zaner, *Problem of Embodiment*, 11.
57. Marcel, *Being and Having*, 11–12.
58. Marcel uses terms body-subject and body-object to describe ways I relate to my body. This will be discussed later in this chapter.
59. Marcel, *Being and Having*, 148.
60. Marcel, *Mystery of Being*, vol. I, 109.
61. Ibid., 100.
62. Ibid.
63. Ibid., 101.
64. Marcel, *Creative Fidelity*, 37.
65. Marcel, *Mystery of Being*, vol. I, 101.
66. Marcel, *Metaphysical Journal*, 274.
67. Marcel, *Being and Having*, 159. Marcel explores the meaning of the *qui* in relation to *quid*, "Any assertion about having seems to be somehow built on the model of a kind of prototypical statement, where the *qui* is no other than *myself*." One's relation to/as one's body is a prototype for how one has at all.
68. James Collins, "Introduction to the Torchbook Edition," in *Being and Having: An Existentialist Diary*, translated by Katharine Farrer (New York: Harper and Row, 1965), xiv.
69. Ibid.
70. Marcel, *Mystery of Being*, vol. I, 100.
71. Ibid., 11.
72. Ibid., 148.
73. Ibid., 82.
74. See Zaner, *Problem of Embodiment*, 55, and Marcel, *Mystery of Being*, vol. I, 99.
75. Marcel, *Mystery of Being*, vol. I, 100.
76. Gabriel Marcel, "Existence and Objectivity," translated by Bernard Wall, reprint in *Metaphysical Journal* (Chicago: Henry Regnery Company, 1952), 333.
77. Marcel, "Existence and Objectivity," 334–335.
78. Marcel, "Existence and Objectivity," 332.
79. Marcel, *Mystery of Being*, vol. I, 92.
80. Marcel, "Existence and Objectivity," 333.
81. Marcel, *Being and Having*, 164.
82. Ibid., 82.
83. Ibid., 83.
84. Ibid., 153.
85. Marcel, *Mystery of Being*, vol. I, 209.
86. Marcel, *Homo Viator*, 15.
87. Kenneth T. Gallagher, *The Philosophy of Gabriel Marcel* (New York: Fordham University Press, 1962), 55.
88. Marcel, *Mystery of Being*, vol. I, 101.
89. Marcel, *Creative Fidelity*, 21.
90. Ibid.
91. Marcel, *Creative Fidelity*, 29. Marcel, *Mystery of Being*, vol. I, 118.
92. Marcel, *Being and Having*, 82, 160.
93. Ibid., 160.
94. Ibid., 149.
95. Ibid., 161.
96. Marcel, *Creative Fidelity*, 17.
97. Gabriel Marcel, *Man Against Mass Society*, translated by G. S. Fraser (Chicago: Henry Regnery Company, 2008), 44.

98. Gabriel Marcel, *Tragic Wisdom and Beyond*, translated by Stephen Jolin and Peter McCormick, edited by John Wild (Evanston: Northwestern University Press, 1973), 38. Marcel states, "there can be no freedom without the possibility of position."
99. Marcel, *Creative Fidelity*, 21.
100. Claude Shannon and Warren Weaver, *A Mathematical Model of Communication* (University of Illinois Press, 1949). Shannon and Weaver describe communication as a process of transmitting information between transmitter and receiver.
101. Ramsey Eric Ramsey. *The Long Path to Nearness: A Contribution to a Corporeal Philosophy of Communication and the Groundwork for an Ethics of Relief.* (New York: Humanity Press, 1998), 1–2.
102. Frank Macke, "Intrapersonal Communicology," *Atlantic Journal of Communication*, 16 (2008), 122–148, 135. Macke offers this description of communication from a communicology approach.
103. Paul Watzlawick, Janet Beavin Bavelas, and Don D. Jackson, *Pragmatics of Human Communication: A Study of Interactional Patterns, Pathologies, and Paradoxes* (New York: W.W. Norton and Company, 1962), 51.
104. Marcel, *Mystery of Being*, vol. I, 90.
105. Marcel, *Mystery of Being*, vol. I, 111.
106. Straus and Machado, "Marcel's Incarnate Being," 133. Straus and Machado offer these as examples of moments in which one feels a heightened sense of value of being alive. This heightened sense of life's value becomes a Marcellian admiration for existence.
107. Marcel, *Creative Fidelity*, 17.

FIVE

Seeking Communion amid Disconnections

Gabriel Marcel's Intersubjectivity

Although we interact in a myriad of encounters with other persons throughout our day, walking among crowds and chatting frequently with colleagues, we may remain disconnected *from* the others surrounding us. In our ongoing interactions with and alongside others, awareness that one is in position to another, communicative attitudes explored in Chapter Four, may be an individual awareness, a solitary experience. Amid our many interactions, we may not hear, not recognize, not encounter, not connect with the other person. The missed connections or disconnections may indicate the absence of sensed presence of the other person, "breaks in presence."[1] Our ongoing disconnections from others reveal a lack of awareness that we are *with* others. These interpersonal disconnections invite our acknowledging the existing gaps between us and seeking ways across these gaps.

We begin to build these bridges between us when we become aware of the presence of others. We experience presence as physical, self, and social presence.[2] We bear a "feeling" of presence, something that acts upon our consciousness.[3] We may also share this awareness of the presence of the other, as sensed copresence.[4] One becomes aware, through an "embodied cognition," that the other is spatially present *with* and attentive to who is sharing hereness and immediacy.[5] Philosopher Eleanor Stump defines "a person *is* present to another person" when one has "direct and unmediated causal contact with and cognitive access to another" and there is "shared attention."[6] When I sense that or am conscious of and focus on the other who is with me, I am present with them.[7]

When I am not focused on the person with me, when I withdraw into my personal, internal world of the mind, I am absent from the other.[8] Absent from the other, I do not sense nor recognize the presence of the other, I cannot acknowledge the presence of the other.

From his earliest writings, Gabriel Marcel offers ways to build bridges across our interpersonal disconnections. Marcel identifies how many times persons do not acknowledge the presence of others. Marcel suggests this failure to acknowledge presence stems from not hearing the call the other's presence makes upon oneself or from not recognizing the presence of another person. Recognizing presence is also an acknowledgment of being, "Being is granted to me as a presence or as a being (it comes to the same for he is not a being for me unless he is a presence)."[9] I, aware that "I am," become aware of another "I am" who is *with* me. This awareness is an act of seeking, recognizing, and/or acknowledging that is both a passive receiving of the presence of the other and an active grasping of the presence of the other. When one lacks this acknowledgment of the presence of the other person, one assumes attitudes of distancing from the other person. Marcel describes these distancing attitudes as positions that privilege self over other. One intersubjective attitude addresses the other by an "I-he" in distinction from an "I-thou" relation. Another intersubjective attitude of *indisponibilité* has no room for the other person in distinction from an attitude of *disponibilité* that makes room for the other. These many disconnections and attitudes permeate our intersubjective interactions and conversations. Marcel proposes that a person is free to cross the gaps between us as persons by choosing a distinct positioning toward the other person. This distinct positioning may refresh interpersonal connections.[10] When one listens for and recognizes the presence of others, when one chooses to address the other as "thou" and to make room for the other, one positions oneself for refreshing interpersonal connections, for possible communion.

Marcel did not always focus his philosophy on these concrete relations between persons initially he interpreted the world with an idealist lens.[11] During Great War of 1914, Marcel, unable to serve in armed forces due to ill health, worked in the Red Cross office of missing persons.[12] In this office, Marcel met daily with persons suffering because they were uncertain if their missing loved ones were alive or not.[13] These terrifying uncertainties about the absence of loved ones impacted Marcel's life-long commitment to attending to the other person's feelings and real experiences.[14] In this work of discerning presence amid absence, Marcel made his turn to the concrete, the lived experience. After the Great War, Marcel committed to a philosophy attentive to who and what is happening here and now.[15] In his own life, Marcel was a person who notably reached out to connect with others on a personal level. Many of the scholars who reflect on Marcel's philosophy also mention personal encounters with Marcel.[16] These personal encounters with Marcel about his ideas add interpersonal dimensions to the ideas Marcel promoted. What began as Marcel's solitary personal search for the meaning of being became an ongoing encounter with intersubjectivity in real, concrete being *with* other persons along the way.

Gabriel Marcel's philosophy which explores being in the world is *not* a solo expedition, a setting off to discover what it means that *I* think about being *with* others. Although secondary reflection involves a personal pausing to recollect oneself in thought, in this reflection, one realizes in this reflection that I am always already *with*.[17] In Chapter Three, when one thinks about being, one experiences being as an invocation upon, a calling made upon me. In Chapter Four, when one reflects on incarnate being as *how* one is in the world, one senses that one's point of reference as body signals one is other to others.[18] Our relations in and of being in the world always sound as invocations upon. As one engages philosophy as a reflection on being, one is "alert for a certain music that arises from its own inner nature if [philosophy] is succeeding in carrying out its task."[19] As one awakens to being, one awakens to being *with:* as ontological exigence, as incarnate being, and as being *with* other people, intersubjectivity.

Intersubjectivity involves an exposure to the other, happening in the presence of the other as other.[20] Marcel's concept of intersubjectivity integrates reflections on intrapersonal and interpersonal experiences. Marcel's philosophy centers on this study of "*beings* taken in their individuality but also affected by the mysterious relations which link them together."[21] This chapter explores how one becomes aware of and seeks communion in these "mysterious relations which link them together." First, this chapter outlines the ways that intersubjectivity make invocations upon us. Second, Marcel's ways of recognizing presence are summarized. While one's acknowledgment of the other person involves hearing and seeing past disconnections, an intersubjective attitude also includes positioning of oneself toward or away from the other person. In the third and fourth sections, this chapter describes Marcel's terms for intersubjective attitudes as an I-thou address and *disponibilité* or *indisponibilité*. Although Marcel does not offer extensive discussions specifically about communication,[22] his philosophy of intersubjectivity as invocations upon and positions *with* an other person contributes significant insight to the meaning of communication.[23] Finally, this chapter explores, how does Marcel's description of communion of presence amplify the possibilities for communication? Awakening to the presence of the other person who is *with* us, positions us to being *with* and *for* the other person, and opens us to seeking more than communication, to being in communion with.

INVOCATIONS TO RECOGNIZE INTERSUBJECTIVITY AS NEXUS OF OUR EXISTENCE

One may physically *be* alongside other persons and not perceive that one *is with* an other person. Marcel contended that only when one "recognized the deep, individual quality of somebody . . . true intersubjectivity arises."[24] The individual quality, or the presence of an other makes "invocations" upon the other person.[25] Hearing an invocation happens as recognizing the other as being somebody. One hears the presence of the other and is aware that one is with or interacting with an other. One is woven in relationship with others from the beginning of our existence, we experience invocations to hear the other who is here with us. Marcel identifies calls to recognize intersubjectivity sounding throughout life experiences and in the foundational experience of existence.

Calls to recognize intersubjectivity sound from one's birth. One is in the world not because one imagines that world to exist, but one is in a world that shapes who one is. Marcel describes how one is a being in a situation,[26] being *in* the world in its most foundational ways. One does understand one's world, through mental acts of representation, one becomes aware of being in the world, "as in a womb."[27] One's first acts in relation in world are receptive acts: receiving from mother's nutrients and growing in response to this nurturing relation. In studying life, my life as incarnate being, one cannot just attend to a solely personal experience, for no life is "just mine," and all life involves "togetherness."[28] In a bodily sense, one is born into a community and comes into existence in the presence of others, "I am the incarnate reply to the reciprocal appeal between two beings."[29] Birth into existence is itself a reply to others, an "I am" for and before the other always as "a given person."[30] One is born in relation to parents, persons, and communities. One receives sustenance, protection, and language from and is dependent upon whomever cares for us. In acknowledging that "I exist," one becomes aware of oneself as being given, as an "I" directed to an interlocutor.

For Marcel, one's sense of self signals the presence of others.[31] A child develops a sense of self through biological or social relations to people who teach the child to express an inner world through communication. The child develops a sense of self in response to being recognized by others.[32] One's sense that "I am unique" emerges as being *distinct from* others, or

> To think of somebody else is in a manner to affirm myself in face of this somebody else. To put it more accurately, the other is on the far side of the chasm, and there is no isthmus between us. But this chasm or separation is something that I only realize if I stop and stand outside myself—picture myself.[33]

Thus, "somebody else" reveals to the reflecting subject a distance between the other and the subject. The other as separate from the subject reveals one's *self* as always in position with an other person. Even the most intrapersonal experience of self bears an interpersonal dimension. This chasm between persons becomes known to me when I reflect upon myself while reflecting from some position outside myself, looking in at myself. One who engages Marcel's secondary reflection as a distancing from may then realize how the other is other to myself. One develops a sense of self in the presence of, and thus, in relation to, others. Brian Treanor offers an excellent description of Marcel's use of the terms self as connected to an experience of the other.[34] For Marcel, in identifying my self, I identify my self *from* others.[35] Even if one pulls back from others, one is still in relation to the others—alienated from them. Even in identifying my own realm of being, I hear this realm told to me *as* who I am, by those who are in relation with me.

An "I am" never exists in isolation as a *cogito* or as a self-sufficient existent. Marcel begins this study of "Who am I?" by attending to an "I am" who is always already in relation to others. In this reflection, "it is not enough to say that it is a metaphysic of being; it is a metaphysic of *we are* as opposed to a metaphysic of *I think.*"[36] Marcel sees that at the heart of my existence, who I am, is not I think, therefore I am, but 'we *are*' therefore 'I am.' The "I am" "in its intrinsic structure subjectivity is already, and in the most profound sense, genuinely intersubjective."[37] Subjectivity, the capacity to be and say I, is born in reply to other beings, as intersubjectivity.

Marcel claims that a "metaphysic of *we are*" forms the "intersubjective nexus" of human existence.[38] In making a claim about a "metaphysic of *we are,*" Marcel engages a particular approach to metaphysics and ontology. Western metaphysics is connected to the study of being as substantive or as a "spatiotemporal entity."[39] In distinction, Marcel suggests, metaphysics is "the activity by which we define an uneasiness and manage partially (and moreover, mysteriously) if not to remove it at least to transpose and transmute it."[40] Dwayne Tunstall argues that Marcel's meaning of ontology does not mean categorization of beings, but rather, Marcel's ontology reflects on "the meaningfulness of those phenomena that enable us to participate in being."[41] How do we participate in being? I participate in being as ontological exigence, incarnate being, as being in relation to others. What could these phenomena mean? Marcel suggests that ontological exigence, incarnate being, and being *with* share a "metaphysic of *we are.*" Thus, how I am in the world, as ontological exigence, as incarnate being, and as being *with* communicate that being in the world is rooted in a relation of "we are."

With what certainty does Marcel make this claim that being in the world is founded on a "we are"? Marcel's philosophy of intersubjectivity has been compared to Edmund Husserl's explorations of intersubjectivity.[42] Paul Ricœur offers a thorough differentiation between Husserl's and Marcel's starting points, methods, and insights into what it means to be in the world and to be *with* others.[43] To make this claim about how existence is founded on a "we are," Marcel assumes a starting point of recognizing the "undeniable presence of the other."[44] One cannot make an *a priori* claim about intersubjective nexus because one is both studying and being studied in this reflection, one is involved in the very assertion.[45] Marcel suggests this intersubjective nexus constitutes how one has any experience of "given."[46] Our intersubjective nexus constitutes a condition for how one relates at all and so one cannot distance oneself to give an objective definition of that which is given. Marcel claims that we bear intuitions about being *with* others. Although one cannot logically reason to grasp this assertion, one does bear "a more or less distinct consciousness of the underlying unity which ties me to other beings of whose reality I already have a preliminary notion."[47] One senses a "notion" of being tied to other beings. This notion of being, at the heart of experience, as *we are,* is experienced as a mystery, known and understood in an indirect way. Both my "experience" and "intuition" point to intersubjectivity being at the heart of existence.[48] Paul Ricœur noted that Marcel offers us this significant insight: "[T]he recognition of the other is not a second step preceded by the certitude of the *cogito,* but rather communication is constitutive of my very existence."[49] At the heart of being in the world, I am always already in the presence of other—I am *with.* One cannot *assert* the presence of an intersubjective nexus at the heart of existence; rather, "[I]t can only be acknowledged."[50] Marcel proposed that through reflection, secondary reflection, one recognizes or awakens to this reality of intersubjectivity that constitutes one's existence.

Marcel calls for attending to human experience, for listening to what we hear in this experience. All human experience begins in birth which is always birth in reply to, in relation to others. Then as we grow as human persons, we develop our sense of self over and against the other who is not me. Marcel hears in one's very relation to being, a relation that is always already intersubjective, I am *with* others. Hearing intersubjectivity at the nexus of existence is an act of acknowledgment that emerges when one listens to one's being as ontological exigence and as incarnate being. Calls to recognize intersubjectivity sound from our first experiences of existence. Learning to recognize this sound of intersubjectivity happens as one awakens to invocations of being *with.*

BEING *WITH*: RECOGNIZING INVOCATIONS OF PRESENCE

The presence of others calls out to be heard. As incarnate beings we experience and are exposed to presence of and as others. Not everyone recognizes what they are hearing in the experience of being with the other person. Marcel describes being *with* as sensing the presence of others, acknowledging that one is *with* others, and awakening to intimate bonds created in being *with*.

What does it mean to experience presence? Presence experienced as body is intimately connected to being in the world. As explored in Chapter Four, reflection, presence as "Here I am!" and as *my* body announces my existence. Presence signals more than "just being there."[51] An object physically *there* with me is not present to me the same way being is present to me.[52] One perceives the other as being before or with oneself. Dwayne A. Tunstall notes that, for Marcel, this is more than a perception of the other who is there, through secondary reflection[53] one "appreciate[es] being" as an "nonobjectifiable presence."[54] Marcel's description of being as "over-abounding" thought, means being is "uncharacterizable" and thus one appreciates being—or presence—as a standing before or a feeling within. Recognition of being happens as a receptive act: "Being is granted to me as a presence or as a being."[55] Because one bears a "sense of existing, of being in the world," one can recognize presence.[56] Recognition of some phenomena means knowing and feeling that this phenomenon is already known, is familiar. If one bears a sense about what being means, one is more open to recognizing the mystery of being in the presence of an other.

Conscious recognition of presence involves an acknowledgment that one is with something *more*, or some*one* more. People, who may be "actual presences" in one's daily life, are "rarely consciously experienced by us *as* presences, we get used to them, they become almost part of the furniture."[57] Interacting in an office with co-workers who are actually there, I may not acknowledge individual persons as being present *with* me in the workspace. Recognition of presence involves a conscious acknowledgment of the other *as* a person.[58]

Because a person exists, he is in the world, *as* incarnate being and as being *with* other persons. As being in a situation, one is always "exposed to" or "permeable" to the presence of others.[59] One's fundamental situation as incarnate being reveals aspects about always already being *with* others.[60] Marcel attempted to describe how we recognize presence by using the spatial terms internal and external. One experiences presence in an internal and external relation:

> When I say that a being is granted to me as a presence or as a being (it comes to the same for he is not a being for me unless he is a presence), this means that I am unable to treat him as if he were merely placed in front of me; between him and me there arises a relationship, which in a sense, surpasses my awareness of him; he is not only before me, he is also within me—or rather, these categories are transcended, they have no longer any meaning.[61]

Marcel uses the concept of before me (outside me, external to me) and within me (inside, internal as me). Being is granted to me, or as Marcel describes it in another phrase from this same work, presence "is a kind of influx."[62] Richard Zaner suggests that Marcel's use of spatial descriptors "outside" and "inside" does not add to the clarity of understanding how one experiences being as incarnate being, as body.[63] When thinking about phenomena like presence, one comes into-the-presence of being, one stands before being.[64] Yes, one is exposed to presence of the other person, but recognizing presence is more than just sensing that the other is there. Marcel seeks to avoid a reduction of the presence of an other to some object that my mind grasps. Words like *before, among,* or *within* fall short of describing how one experiences the presence of an other. Thus, acknowledging the being of the other who is *with* reveals an intimate bond that "arises" between us, a bond of based on mutual recognition.

Leaving aside his internal-external distinctions, Marcel offers a better description when he relates *experiencing* presence to participation.[65] One cannot objectively—or as a distanced analysis—identify presence. Presence "can only be invoked or evoked."[66] One actively is present to the other and one is called forth by presence. Persons who recognize presence of the other, participate in being *with.*[67] One who participates in being *with* is not the only actor who evaluates if the other person deserves recognition or not. Because presence denotes being, human beings are present even if others don't recognize them as beings. Presence can be ignored but not erased. Being cannot be un-been. Because being is really a mystery, we can be among beings even if we do not realize it, because this or that incarnate being always is, even if no one acknowledges her presence, she is. Incarnate being over-abounds any one's realization *that* someone is there.

One sometimes uses the word *with* to describe relations that are not—for Marcel-- actually being *with.*[68] Marcel reserved the phrase being *with* for being aware that one is present with another person. People might be seated with other passengers on a train and no relational dimension exists in this positioning.[69] A relation of being *beside* indicates a merely physical, spatial relationship between a subject and one who is related to as an object. When persons share *more than* a physical space, they become united in some way, in a relation beyond a subject-object relation, they are *with* one another.[70] Between the two persons, "there must be an appeal, an invocation, an 'abide with me' that is more or less clearly enunciated."[71] Marcel connects invocation to presence and to being *with.* This description of being as invocation was foundational to Marcel's description of intersubjectivity.[72] The presence of the other is heard as an invocation upon and one may recognize this invocation as indicator that I am *with.*

Marcel describes this sense that the other is *with* me as bearing an affective dimension. We face a temptation to identify the presence of the other by only using our mind: this being is an object that is present *to* me.[73] In interpersonal encounters one may "feel[] like one is not *with* ("avec") the others, that he is being excluded from a certain community to which he feels he rightly belongs."[74] When we are not extended this acknowledgment of being *with* the other, we feel that we do not belong. Yet, Marcel suggests if we listen to experience, which also includes a vital dimension, we bear a feeling about being *with* an other who is with me. Beyond sensory certainties that I am with someone, one may hear this internal voice:

> Even if I cannot see you, if I cannot touch you, I feel that you are with me; it would be a denial of you not to be assured of this. *With* me: note the metaphysical value of this word, so rarely recognized by philosophers,[75] which corresponds neither to a relationship of inherence or immanence nor to a relationship of exteriority.[76]

Marcel describes a relationship that is distinct from physical seeing and touching. One *feels* that the other is *with* oneself. To deny that "you are with me," even if out of sight and touch, is to deny that you are. This being *with* relationship is not a relationship of immanence which is transmitted to the subject from external sources. Before this quote in "Ontological Mystery," Marcel describes how one senses actual presence, the other person "is not only before me, he is also within me—or, rather, these categories are transcended, they have no longer any meaning."[77] My awareness that the other is *with* me is an internal, ontological awareness. One's sense of being *with* springs up from within bearing affective dimensions and is in response to actual presence, to incarnate being.

Being *with* another person means more than being physically located beside or exposed to the other. Being *with* signals an internal, affective relation between at least two persons. When I become aware of a relation between, "I restore, in a manner, between him and myself, a community, an intimacy, a *with*."[78] Marcel also uses "intimate" to describe a relation of indivisible unity in how I relate to my body.[79] Intimacy indicates a bond of affection between, a mutual sharing. Intimacy insinuates vulnerability in the sharing of oneself *with* the other person. *Both* persons have heard an "abide with me" and become aware of the other being *with*. When both persons are aware of being *with*, an opening appears for "genuine *co-esse* . . . that is to say, genuine intimacy."[80] Genuine relations of intimacy emerge in mutual being *with*.

Marcel connects co-being with an intimate relation between. *Co-esse* or being copresent involves an awareness that the person is more-than just physically beside me.[81] The presence of the other person makes an invocation upon another person, "an 'abide with me.'"[82] Copresence is sensed as a "mysterious *co-esse*."[83] In another reflection, Marcel elaborates, "*With* me . . . is of the essence of genuine *coesse*."[84] Hearing the invocation of the other and being aware that one is *with* the other, lays the ground for the possibility of *co-esse*. Marcel also suggest "the English noun *togetherness*" best explains how "between you and me," a shared sense of being *with* can emerge.[85] Many times we are surrounded by people who are "very rarely consciously experienced by us *as* presences; in so far as we get used to them."[86] If one engages secondary reflection, one may grasp what that appeal of presence means, and recognize another incarnate being like oneself who is *with* one.

Because presence "can only be invoked or evoked," the presence of the other person bears a type of "borrowed otherness."[87] The presence of the other makes an appeal upon me, will you acknowledge *that* I am? All the time, one is exposed to, permeable to, the presence of others. Awakening to being *with* this person means hearing the appeal of presence of the other person, recognizing the presence of this person who is with me, and sometimes being recognized by the other person, as also being *with* me. Being *with* the other in this deeper way also shapes how one addresses the other person. One's position toward the other person is reflected in how one addresses the other.

INTERSUBJECTIVE ATTITUDES: ADDRESSING THE OTHER AS I-THOU

Marcel suggests that in interactions with others, one's choice of pronoun for the other person reflects one's attitude toward the other person. A person relates to the other person as a *he* or as a *thou*.[88] Marcel's attentiveness to an I-thou relation joined conversations of his historical moment. Although Martin Buber would be better remembered for his discussion on I-Thou relations published in 1923, already in 1915 in his *Metaphysical Journal,* Marcel was mentioning the difference between an I and thou and an I and he.[89]

One's form of address of the other person, as I-he or I-thou, reflects a way of relating in attitude and in position toward the other person. Some possible attitudes toward the other person may be respect and esteem or disregard and contempt. One's attitudes inform one's position toward the other person. In interactions with others, one assumes an address of I-he or I-thou as ground for encounters that may become intersubjective meetings.

A communicator may select a third person (I-he) or second person (I-thou) form of address. In an I-he dynamic, the other is a "somebody, 'that person' in my eyes."[90] Marcel uses a scenario about self-centered person at a party to exhibit the different approaches. When someone approaches the self-centered person and tries to converse, the self-centered one thinks to himself, "Why is *he* talking to me? What is *he* after?" distancing himself from the other person preventing "a genuine encounter or conversation *with* him."[91] The self-centered person views himself on a different level than the other person. Mentally addressing the other as infringing upon his world, there is no room for an encounter with. In treating the other as third person, one deals with the other as object, as if he were absent—not physically there or not genuinely attending to the conversation.[92] This I-he address indicates that I view the other as less than a fully autonomous subject who can be encountered.[93] Addressing the other "in the *third person,* I treat him as independent—as absent—as separate. . . . I define him implicitly as external to a dialogue that is taking place, which may be a dialogue with myself."[94] A possibility for interaction becomes a conversation by one, a monologue.

Marcel countered this approach to the other as he with an approach to the other as thou. In an "open" attitude of I-thou toward the other, one sees the "other *qua* other," as irreducibly other.[95] Seeing the other as thou, "I treat him and apprehend him *qua* freedom."[96] The other in his freedom cannot be reduced beyond his own autonomy nor beyond his own free actions, and the other cannot be confined to my idea of him. Marcel describes this attitude as being "open to him, in so far as he is a Thou."[97]

Respecting an other *qua* other indicates respecting the other's freedom. One of the aspects of being a person includes having choices about how one acts. Freedom involves choice of differing attitudes and ways of positioning oneself toward the other. One is free to treat the other person as object or as thou. Marcel wrote, "When one "treat[s] the other as 'Thou', [one] treat[s] him and apprehend[s] him *qua* freedom."[98] Treating the other as thou means deferring to the other the freedom to act and be in the world as he pleases. In our ways of addressing the other as thou, with the respect, deference, and esteem they deserve, we "help him, in a sense, to be freed, [they] collaborate with his freedom."[99] My address of the other, as I-thou, my position toward the other of respect collaborate with his ways of being free in the world. My personal living of freedom is impacted by how others view and address me. My freedom to be in the world is empowered in being addressed as a thou.

Marcel differentiates types of interactions with others, based on how one views the other person as a "he" or a "thou." Marcel describes a scenario of a person sitting next to a stranger on the train. If the two just talked about weather or news, the two persons relate to each other as "he," merely communicating with "signs which coincide with signs of mine."[100] One understands the words but misses something. The more that the stranger spoke to the person, sharing not about the weather nor about war news but about him own personal history, something happened in the relation. As the person became aware of the stranger as "somebody" particular, he became aware of himself as "somebody else."[101] The stranger becomes "less and less an object for me," and "his or her presence is such that I am less and less *him* for myself—my interior defenses fall at the same time as the barriers that separate me from somebody else."[102] Marcel returns to this same example on the train, in his

1933 reflection, *Creative Fidelity*. When the two communicators realize they share some common interest, their "outer defenses fall at the same time as the walls separating [them] from the other person fall."[103] A connection happens, or a "bond of feeling" emerges between the communicators, and the two "become *we,* and this means that he ceases to be *him* and becomes thou."[104] When one senses the presence of the other, the other becomes a thou. One shifts from sharing news in the same language with "somebody" (I-he) on the train to sharing aspects of oneself in the presence of someone one who is "somebody else" (I-thou).

In situations in which the other seems or is absent, one experiences an I-he address, and monologue ensues. If, in the interaction, one addresses the other as I-thou, the other as other is respected in their freedom. If both participants sense the other as being *with,* as being a thou, bonds between may emerge. Intersubjectivity arises in environments of respectful addresses, of respect for irreducibility. Changes in addresses from I-he to I-thou include changes in position in relation to the other. Judgments based on disregard and contempt give the one who judges the false sense that he is better than the other, that the other is less-than. This being less-than may translate as being less free, less valuable, less enlightened, less insightful, or less worthy. This sense that the other is less-than is false. No person is less-than. An address of I-thou counters this false sense of inequality. Marcel reminds us that the other *as* freedom is to be treated with an I-thou address. The other as free person, autonomous, incarnate being is irreducible to my sweeping generalizations or categories. Marcel calls for refusing the false narrative that allows an I to set himself up above any other person and address them through I-he. This "spirit of abstraction" lessens—in the mind of one who abstracts them—the person to one who is less than irreducibly free, one who is respected, and presence that cannot be revoked. An address of I-thou respects the other as free person, as deserving of all respect, and as irrevocable presence. This way of addressing the other is connected to one's attitude toward the other, Marcel names these intersubjective attitudes *disponibilité* and *indisponibilité*.

INTERSUBJECTIVE ATTITUDES: *DISPONIBILITÉ* AND *INDISPONIBILITÉ*

Marcel used the French words *disponibilité* and *indisponibilité* to characterize ways one is in relation with others in the world. In 1930, Marcel first introduced his philosophical concept of *disponibilité* and continued to reflect on the concept throughout his work.[105] *Disponibilité* loosely translates into English as availability or disposability and *indisponibilité* translates as unavailability or non-disposability.[106] Marcel, bilingual in French and English, suggested that the English term availability did not fully describe what he meant by the philosophical phrase *disponibilité*.[107] In his lectures in English, he would frequently use the French word *disponibilité* and describe what he meant by the phrase. This project follows other scholars in primarily using Marcel's terms *disponibilité/indisponibilité* and using the French adjectives *disponible/indisponible.*[108]

Disponibilité/indisponibilité denote ways of being with others. Joe McCown, in his detailed description *Availability: Gabriel Marcel and the Phenomenology of Human Openness,* suggested that Marcel viewed *disponibilité/indisponibilité* as attitudes, as positioning toward the other person. McCown notes that these attitudes are invoked by awareness/lack of awareness of presence. Paul Ricœur described Marcel's *disponibilité* as "an ontological movement," an attitude toward being that reveals an openness/closedness to the presence of the other.[109] Rodick explains *disponibilité* as a "condition of being accessible through intersubjective encounter."[110] *Disponibilité* describes one's attitude, positioning, ontological movement, and condition of being. This project builds upon this description, proposing that Marcel's relations of *disponibilité* and *indisponibilité* constitute praxes of response in intersubjective meeting.[111] Praxes for intersubjective meeting points to Aristotelian praxis as theory-informed action and encompasses reflective actions carried out by persons in communities.[112] In intersubjective praxes, participants assume positions for or withheld from the other based on how they view themselves and the other person.

Before Marcel identified the intersubjective attitudes *indisponibilité/disponibilité,* he explored the capacity to feel with, or sympathize, with another person. In his 1925 work, Marcel offers two counter approaches to feeling toward a passer-by, the other who is not known to me. One could refuse to reduce the other to object and "sympathise with [the other person] and embrace his 'interior becoming;'" or, one could treat the other as "that other body" and thus, remain closed off from understanding the other's internal world.[113] The distinction between viewing the other as person *with* me or as "that other body" shapes one's position toward the person: an opening to or closing off from the other person. One's being open to the other person will lay the ground for understanding and sympathizing with him. One's being closed off from the other person prevents one from beginning to understand the other person.

A few years after his exploration in 1925, in 1933, Marcel spelled out *disponibilité* and *indisponibilité* as intersubjective attitudes, more than just a one-time interpretation of the other person, these attitudes generate ongoing actions, positions toward, and ways of being in the world. The attitudes *disponibilité* and *indisponibilité* reflect one's general disposition toward other persons: being open or closed to sharing oneself with others. Marcel connects these attitudes toward the other to being able to communicate with the other. Living alongside others, as incarnate beings with incarnate beings, one is in contact with those around him. As noted, a person's "condition as a living being . . . exposes me to, or better, opens me up to, a reality, with which I somehow communicate."[114] As a living being, one is exposed to or in contact with other persons, and seeks ways of communicating with the other. Marcel follows this description of being exposed to others with the question, "On what conditions can I communicate with this other reality?"[115] Marcel answers this question, one must make "room for the other in myself."[116] *Disponibilité* and *indisponibilité,* attitudes for being with others, are connected to exposure to—-being open to or closed off from others—and to having room for the other within oneself. A praxis of *indisponibilité* reveals being who is encumbered by oneself and a praxis of *disponibilité* points to one who has or makes room for others.

Indisponibilité*: Being Encumbered by Oneself*

The *indisponible* one withholds and safeguards oneself, remaining closed off from others, in the name of self-protection. While the *indisponible* person may be physically present, listening to the other person, actually he "gives me nothing, he cannot make room for me in himself, whatever the material favors he is prepared to grant me."[117] *Indisponibilité* indicates being "encumbered with one's own self"; weighed down by one's own concerns; unable to acknowledge any presences beyond one's self-centered realm.[118] Because the *indisponible* person views his life as "quantifiable, as capable of being wasted, exhausted, or dissipated," he is all the more anxious about having his life infringed upon by others.[119] When one carefully measures what one has—quantifiable realities like time, money, good health—one always sees the possibility of losing these resources. When one becomes fixated on keeping one's health or finances in tact or as mine, one's anxiety about possible threats grows "to the point of choking us" and so one builds higher walls of "self-defense" to protect these limited resources.[120] This anxiety and self-concern, signs of *indisponibilité,* reflect a "state of inner inertia," or feeling frozen by fears, pessimism, or a lack of hope.[121] *Indisponibilité* appears as a self-laden neighbor who is closed in by his own protective fences and unable to hear the presence of others.

These barriers around one's inner world, to protect one's own health or fortune, reveal territories of the *indisponible* one who does not engage intersubjective relationships. This person has turned in on himself and become ego centric. If I am "completely absorbed in myself, concentrated on my own sensations, feelings, anxieties, it will obviously be impossible for me to receive, to incorporate into myself, the message of the other."[122] The one who is self-absorbed has no room for others within. "Burdened" by oneself, one "feel[s] an empty void and crav[es] to be confirmed from outside, by another."[123] The *indisponible* one is only oriented toward the other person so as to receive *from* him. He seeks confirmation from the other but does not extend acknowledgement of the other as other. The *indisponible* one assesses the others in terms of their usefulness to him. Self-centered, he views the other as "at my disposal."[124] The *indisponible* person views the other as less-than, as something that serves his purposes. In praxes of *indisponibilité,* one engages some interactions with others, but in a self-serving way as means of maintaining one's focus on oneself. *Indisponibilité* reveals a being who is for himself, while *disponibilité* points to one who is *for* the others.

Because *indisponibilité* reveals barriers between persons, Marcel suggests *indisponibilité* alienates one from the other person. Before a person who is suffering, the *indisponible* one reports, "I feel absolutely nothing."[125] *Indisponibilité,* an incapacity to sympathize with others, "is rooted in some measure of alienation."[126] Alienation from the other person reveals separation from the other person. For Marcel this alienation is a modern "dis-ease," an unease of the modern person who is ill-at-ease in the world due to alienation from self, neighbor, or even from one's own body.[127] Ill-at-ease with others, *indisponible* persons distance themselves from others, creating larger distances that become harder to cross.

Disponibilité: *Having/Making Room For the Other*

In response to our common experience of *indisponibilité* as a self-centered project, Marcel proposes *disponibilité* as a making room *for* others. The *disponible* one is present with and exposed to the other. Marcel gave the example of when someone is in pain or needs someone to talk to, a *disponible* person is "'present'—that is to say, at our disposal."[128] *Disponibilité* means being present to and "at the disposal of" the actual person before me.[129] *Disponibilité* is "the act by which I expose myself to the other person instead of protecting myself from him."[130] One opens oneself and one's inner world to be shared with the other person. The *disponible* one is not "self-contained; on the contrary such a being is open and exposed."[131] In *disponibilité,* one forfeits his "right to dispose freely of himself" and chooses to be *for* the other.[132] One is open to, exposable to the other who is *with* him. A *disponible* person dwells in the presence of the others, on a ground not her own, vulnerable to the presence of the other.

In being at the disposal of the person before him, a *disponible* person is also open to the presence of the other. In *disponibilité* one participates in an "'unveiling' of Being," the recognition of being in the presence of the other.[133] As a listener, he communicates presence through "a look, a smile, an intonation, or a handshake."[134] *Disponibilité* means witnessing to the presence of others. Praxes of *disponibilité* become ground for being *with* the other as praxes of intersubjectivity.

Marcel's describes *disponibilité* as being *for* the others. *Disponibilité* means having "one's resources to hand or at hand."[135] Having one's resources at hand means being able to access/share one's inner resources.

Marcel used a monetary metaphor for extending oneself toward an other person. He distinguished *disponibilité* from *indisponibilité* in terms of credit:

> It will perhaps be made clearer if I say the person who is at my disposal is the one who is capable of being with me with the whole of himself when I am in need; while the one who is not at my disposal seems merely to offer me a temporary loan raised on his resources. For the one I am a presence; for the other I am an object.[136]

A *disponible* one extends himself to the other without conditions or expectations for repayment. The *indisponible* one relates to others expecting remuneration. Marcel uses similar terms in describing relations of an I-thou, one who relates to another person as a thou extends "credit," or trust, toward the other.[137] One believes that the other is trustworthy, and thus one extends "credit" that "is open and unconditional."[138] Unconditional extension of being *for* an other person is an attitude of *disponibilité*. A conditioned being there for others, *indisponibilité* reveals ground that is withheld from the other and a position of reserving for oneself, not fully available for the other.

Disponibilité is an intersubjective praxis of participative receptivity of the other person. Henri Bugbee describes Marcel's term *disponibilité* as a way of being simultaneously active and passive in engaging with others.[139] In *disponibilité*, one is with the other person by blending active and passive engagement. The *disponible* one is receptive in a reflexive and directed-toward-other way. In reflexive *disponibilité*, one remains actively open to "being unconditionally claimed" by the other's needs or presence; and in directed *disponibilité*, one "heeds" or listens attentively to the claims of the other people.[140] A *disponible* person is open to hearing the other as an invitation for response and is ready to extend a response. One participates in *disponibilité* by remaining open to the needs of the other and actively listening for the invocations to be *for* the other.

The *disponible* one and the *indisponible* one seem to relate to others from different starting points. *Indisponibilité* springs from an internal landscape of needing to defend one's own territory and resources. Marcel connects the ability to be *disponible* with feeling at home as oneself. *Disponibilité* involves a receptivity and openness to others that can happen if one has a place in which to receive others. The *indisponible* one is encumbered by concerns about his own resources. Understandably, when one's own resources are threatened, one pulls in to ensure that one's needs are met. Marcel describes *disponibilité* as receiving another person much like welcoming the other into one's home. But one must first feel "at home with self," as oneself, as who one is.[141] Only he who feels at home or at peace with himself, can invite an other person to participate in being at home *with* him.[142] One who does not feel one's resources threatened has "room for the other," has resources at hand, and has ground from which to extend an "act of hospitality," as "gift of self," and as "gift of what is one's own."[143] *Disponibilité* involves an opening of one's self to be shared with the other person.

Marcel suggests that one who is able be *disponible* for the other received this other-centered attitude as a gift. In *disponibilité,* when one extends "credit to another, he is making a gift of himself."[144] One does not give oneself in a transaction of giving so as to get in return or to feel good about giving.[145] The praxes of *disponibilité* for others is an unconditioned offering of gift. The *disponible* one gives, in an act of expansion, opening up to, sharing with, without conditions for receipt of or response to the gift. If one gives so as to satisfy oneself, that which is given is conditioned and not actually gift. *D isponibilité* as an unconditioned giving springs from a spirit of generosity. This generosity appears inexplicably in some people, and so Marcel also calls this generosity a gift.[146] *Disponibilité* springs up from one who has received gifts and who has resources to share from and for others.

Disponibilité and *indisponibilité* are constituted by one's attitude toward the other and one's ability to be *for* the other. Praxes of *indisponibilité* reflect a self-centered focus that alters one's capacity to be *with* or feel for others. The more *indisponible* a person is, the more he blocks himself off from the threat that the other is, the more the distance between people grows, and the ground for intersubjective meeting disappears. In distinction, praxes of *disponibilité* indicate an orientation toward the other that opens up one to be *with* and *for* the people àround her. The *disponible* one, at home with oneself, maintains oneself open, exposed to the other, receptive to the other's needs, with resources at hand to be given unconditionally as she continually makes room for the other. One who makes room *for* the presence of the other opens ground for being *with.* Being *with,* feeling *with,* and being *disponible for* offer openings to encounter fulfilling meetings along the way. As one assumes *disponibilité* as being *for* others in the world, possibilities for communication emerge.

MORE THAN COMMUNICATION: COMMUNION OF PRESENCE

Returning to Marcel's reflection on problems with communication mentioned in Chapter One of this book, we turn to extract what Marcel's insights contribute to our understanding of communication. Marcel differentiates types of communication and how this communication affects intersubjectivity. As discussed in Chapter One and at the end of Chapter Four, I view communication, in a general sense, as how we share what we mean. What I mean may imply what I think, what I understand, what I feel, what I desire, what I imagine, or what I hope for. Sharing involves translating or expressing that meaning into verbal or nonverbal codes.[147]

Communication scholar Ramsey Eric Ramsey suggests a broader definition as communication is "the manner in which persons relate to their social world, to others, and to themselves."[148] Communication involves meaning, expressing, sharing, participating, and relating with ourselves, others, and our world.[149] Although Marcel did not offer explicit definitions for communication, he offers explanations about intersubjective encounters that contribute valuable insight to our understanding of communication. This section explores, how does Marcel's description of communion of presence amplify the possibilities for communication? Marcel's explores the ways interpersonal communication often breaks down and a way that communion of presence may change the interaction. This project suggests Marcel's communion of presence offers a *telos* for communication.

Communication Breaking Down

Gabriel Marcel identifies common life experiences of being uncertain if one is really present *with* an other person. Marcel connects these problems of presence to communication. In his 1950 Gifford Lectures, Marcel describes a typical experience between two persons who, while standing in the same room, fail to be present one with the other. Between the two persons,

> A kind of physical, but merely physical, communication is possible; the image of passing of messages between a reception point and an emission point. Yet something essential is lacking. One might say that what we have with this person, who is in the room, but somehow not really present to us, is communication without communion: unreal communication, in a word. He understands what I say to him, but he does not understand *me:* I may even have the extremely disagreeable feeling that my own words, as he repeats them to me, as he reflects them back to me, have become unrecognizable. By a very singular phenomenon indeed, this stranger interposes himself between me and my own reality, he makes me in some sense also a stranger to myself; I am not really myself while I am with him.[150]

This passage generates significant descriptions of communication. The two persons exchange words and although the message is understood, the person who delivers the message does not feel understood. Communication breaks down in different ways.

Marcel's secondary reflection offers an entry point one to identify what is missing in this scenario. A scientific approach to this situation, suggests that a message was effectively transmitted and communication successful. A scientific study fails to account for intersubjective experiences of being *with*, communicating more than words. To study the persons involved in this scenario, one must allow for more than material, sensory communication. These invocations are heard experientially not as measurable soundwaves. To grasp and understand this type of communicative encounter, one uses secondary reflection to attend to dimensions of the encounter that go beyond the words or gestures. To understand what true communication means, one uses secondary reflection to study dimensions beyond the material, sensory aspects of communication. Marcel's secondary reflection of this experience attends to a communicator who feels misunderstood, alienated, and distanced from the other and reveals the three ways communication has broken down. Marcel describes these breakdowns as merely transmitting messages, lacking feeling of being *with* and distancing, and misunderstanding.

The first way in which communication breaks down happens in the misinterpretation of what real communication is. Marcel describes the verbal exchange of messages from "emission point" to "reception point" as "merely physical" communication. In another reflection, Marcel describes a similar type of communication happening in conversations taking place on a public train. Two strangers can discuss a subject because they share "signs or symbols," a language that allows them to exchange information.[151] This emission-reception communication resembles communication as transmission: one emits information and the other person receives the information.[152] These two persons may share a physical space and may even share ideas in back-and-forth conversation but the two persons have not experienced that the other person has understood one's entire message. Although an exchange of messages has happened, Marcel calls this "unreal communication."

A second way communication falters is the lack of being present *with*. Marcel calls this lack "something essential." To know if the other person is really present with us, we discern her/his presence. One communicates presence using more than words. As noted earlier in this chapter, we may hear the other's presence as a calling upon us. One may sense that the other is present with oneself, called by an "abide with me," to be *with*, to *co-esse*. The *indisponible* one is too preoccupied and weighed down by oneself to hear the invocation. Being present *with* may break down in the being open to the other person, in the hearing of the invocation, or in the acknowledging of being *with*. This lack of presence results in a feeling of distancing between the persons. The person "somehow not really present to us" is absent to his neighbor. One may sense that the other is physically there *with* oneself but actually absent to us. One feels that this person here in shared space is "far further away from us" than someone who is actually thousands of miles away.[153] Absence, felt as lack of presence among persons, creates feeling distant from those who are actually alongside us.

A third way participants in this conversation experience failing communication happens when the communicator feels misunderstood. When the listener is not really present *with* me, he "understands what I say to him, but he does not understand *me*." The listener understands all the words I am saying but does not understand what I mean by those words. I experience a disconnect between what message I am wanting to share with the other person and what message he is understanding. Marcel describes, when the receiver tells me what he thinks I have said, "my own words . . . have become unrecognizable." What I had wanted to communicate is not at all what was communicated, communication has been ineffective. One feels alienated from one's own words uttered moments before. Words spoken to those who are not really present can be misheard and so the communicator feels misunderstood, "he does not understand *me*." One's attempt at communication has led to actual misunderstanding.

Failing to communicate, the persons in this encounter do not share messages, meaning, or experiences. A distance remains between the two persons. Misunderstanding, withholding, and being absent create intrapersonal and interpersonal alienation. Something is missing in the communication.

Communion of Presence

When one experiences misunderstanding or barriers between, one becomes aware that "something is lacking." Marcel suggests that the communicators lack communion between them. In his 1919 diary entry, he first used the word "communion" followed by the parenthetical comment: "communion (I am looking for an equivalent of *Mitsein*)."[154] Returning to his 1950 scenario, Marcel follows the scenario about a lack of communion with a distinct experience. In the same setting, two persons are both in the same room, but this happens: "When somebody's presence does really make itself felt, it can refresh my inner being; it reveals me to myself, it makes me more fully myself than I should be if I were not exposed to its impact."[155] A presence that makes itself felt—sensed by the other person—is given to and acts upon the other person. Even if persons are with one another in a spatial or temporal way, to feel the presence of the other, there must be conscious recognition that the other person is *more than* just being there.

This second scenario offers generative descriptions of communication *with* (meaning incorporating) communion. While the two persons exchange words and are aware that the other is present, the possibilities for communion emerge. Marcel describes these openings for communion of presence as sharing more than words, being aware of being with, forming relational bonds, and feeling understood.

Beyond the objective words exchanged, the two persons come in contact with one another and communicate *more than* the messages they transmit. In his 1919 reflection, Marcel differentiates communion from transmission. Transmission might happen "by letter, telegram, etc." but "communion" happens when one is in "sympathetic communication" with an other person.[156] At a much later date, 1950, in his Gifford Lectures, Marcel again contrasts transmission of messages with "communion of presence."[157] One can use communication-as-transmission to effectively share "purely objective messages."[158] But for "communion in which presences become manifest to each other," one must move beyond transmission of messages to "the threshold of being."[159] When communicators share being *with* each other, an opening for more than communication appears.

A second opening for communion is the shared sense of being *with* the other, of communicating presence with each other. We do not typically experience communion of presence in daily conversations. More commonly, we carry out superficial conversations in which we or the other is semi-present in and semi-absent from the interaction. Then mid-conversation, I realize we share some experience in common, I sense, "you too," and our previous shared "solitude" suddenly becomes a "transforming encounter."[160] We are still sharing words and ideas, but suddenly, the other has "ceased to be someone with whom I converse" and we become a "unity."[161] This type of interaction signals communication of more than words, a communication as communion of presence.

Although Marcel does not offer an explicit, clear definition of what he means by communion, his reflections pieced together point to the meaning of communion. Marcel explains "communion in which presences become manifest to each other."[162] In another reflection on this same scenario, Marcel adds a detail to the experience noting when "it *is he himself saying these words,*" copresence is experienced.[163] When the human person is fully *there* vitally and consciously communicating as "*he himself,*" copresence is possible. When someone's "presence does really make itself felt," a mutual sense of being present with, *co-esse,* may emerge. Participants communicate presence as incarnate being and as being *with.* As incarnate beings who manifests that they are present, they sense that the other person is also present with, being *with* one another. "Real communion," "a reciprocal openness between individuals," is the foundation for intersubjectivity.[164] Communion of persons is experienced as mutual opening to, being *with,* and *co-esse.*

A third opening for communion of presence emerges in the forming of bonds between persons. In his description of two strangers communicating on the train, Marcel notes how at first the two strangers talk about events or the weather. But then they realize they both have visited the same place, love the same book, or appreciate the same things. The other person, once a "he" becomes a "thou," a "unity" emerges when in the presence of the other, I and thou become a "we."[165] A bond, on a level of feeling or affection, between them emerges and how they look at the other person has changed. This realm between, an "indistinctness of the I and the thou," is a "region of fructifying obscurity transcending the closed systems in which thought imprisons us, where beings may communicate, where they *are* in and by the very act of communication."[166] Marcel suggests that in moments like this, experiencing communion with another person, we have transcended our individual thought process that is closed off from others, and we find ourselves in a mysterious realm between us, in communication by being *with* the other person. Communion, a bonding, sharing, communication between two persons, creates a mysterious region beyond or deeper than our individual worlds.[167] Marcel suggests that "meeting, *Begegnung*" offers a more active word for describing the type of intersubjective relation emerging in this encounter.[168] Both persons meet as "being to being" sharing an appreciation that binds them in a relationship together.

Marcel proposes that communion of presence forms ground for community. Every experience of being *with* an other person is a meeting and "within the meeting there is created a certain community."[169] When we have gone beyond polite conversation about the weather, when we have become thou for the other, we experience a "a co-belonging."[170] I feel that I belong in a community when I feel included, a part of the group, and aware that I am at home with them. When we talk with people about superficial events, we maintain polite chat about anything that is not personal, as we talk without really revealing anything about ourselves. Marcel contrasts this superficial encounter with the intimate sharing that happens in communion of presence. When my presence is acknowledged, my inner being is refreshed. When I acknowledge your presence, your inner being is refreshed. Being recognized for who I am, I feel welcomed because I am, and you are *with* me no matter how I am. Marcel speaks about this as *a receiving* the acknowledgment of presence and as *an extending* the acknowledgment of presence. The "more my existence takes on the character of including others . . .the more I am."[171] One who recognizes presence of others, communicates to the others, "you belong" and "you belong *with* me." When we mutually acknowledge the presence of the others, that they belong, our being*s* are refreshed.

In a fourth way, this communion of presence with the other allows one to understand oneself. The more one understands oneself, the more *disponible* one is for communication with others. Experiencing the presence of the other person "reveals me to myself."[172] Being with others uncovers aspects of what it means to be human. In the counter example, when one feels alone in an exchange, one feels that one's words do not express what one truly means. When in communion of presence *with* an other, I understand who I am, what I am trying to express, and that the other gets what I am trying to express. Marcel addresses a key aspect of communication: when one can communicate about one's inner world and one feels that one's communication is understood, one better understands oneself. Being with someone who validates *me* as who I am, as a *thou* who *is* free,[173] I, myself, am enriched by this encounter. This communion "makes me more fully myself than I should be if I were not exposed to its impact." Marcel identifies at the heart of human existence: who I am and my longing for being. Being in communion *with* another person leads one to better understand oneself and to feel more fully oneself.

An intersubjective encounter that communicates *more than* words unites us, bonds us, and is self-revelatory. Communication that is *more than* transmission of messages opens participants to a communion of presence. In his 1940 text, Marcel offers an expanded description of communication that approaches communion of presence,

> Literally speaking, we communicate; and this means that the other person ceases to be for me someone with whom I converse, he ceases to intervene between me and myself; this self with whom I had coalesced in order to observe and judge him, while yet remaining separate, has fused into the living unity he now forms with me. The path leading from dialectic to love has now been opened.[174]

Marcel's description of communication sounds in distinction from the communication breakdown experienced in the earlier scenario. When I "coalesce" or objectify the other person with whom I converse, I stand at a distance from them to observe and judge. But then some change may happen, a change in the way they relate to one another, and the communicators begin to share more than words. After the change, I cease to feel alienated from my own words, and now I feel that we share something together. The other person is not the object *at* whom I speak but a copresence *with* whom I *am*. Both communicators have formed a "living unity," a communion of presence. Beyond our back-and-forth discussion of ideas, a new path to appreciation of being opens. In this affection, in this appreciation, love for and from the other becomes possible. In this passage, Marcel names communication as bearing the possibility for communion of presence.

Communion of Presence Is Telos *of Communication*

Marcel identifies "unreal communication" as "communication without communion." As noted in Chapter Four, Marcel speaks these words in 1950, at a time just after Shannon and Weaver proposed communication as process of transmitting information.[175] Since the 1950s, the discussion of the meaning of communication has markedly altered, expanded and transformed.[176] Marcel's description of communion of presence uncovers insightful dimensions of communication. This project explores, how does Marcel's description of communion of presence amplify the possibilities for communication?

Communion of presence offers a *telos* of and for communication. A *telos*[177] of communication is that which is the fulfilment of communicating. *Telos* is not interpreted merely as a static end or goal at which one hopes to arrive in the future nor as an imaginary future that is ever elusive.[178] A *telos* for communication points beyond what is happening now to a not yet here future. A *telos* of communication involves a deeper experience of communication as that which coincides with what communication was, is, or could be. A *telos,* as constitutive to, shapes my communication since my beginnings and calls me to envision where we are headed in our future communication. This project attends to communication taking place, or not, in our moments of interpersonal encounter. Communion of presence offers a *telos* for communicating in everyday encounters between strangers, friends, colleagues, or loved ones.

In our everyday interactions, we frequently fail at feeling heard, accompanied, or understood. Although communicators may effectively transmit their messages, participants often feel alone, distanced from the other, or misunderstood. When we experience communion of presence, our communication alters completely, becoming a fulfilling experience.

We initiate communication with the other so as to express something to them. My message as words and symbols may be transmitted while I still feel misunderstood. Someone could grasp the content of my message and not have grasped my situation. Marcel describes comprehension of objects (the words of message) as distinct from understanding the person (the one who gives the message). We do not understand the other person through an act of "apprehension of intelligence."[179] The other person, as incarnate being and as presence, makes an invocation upon us. I understand the other person, who is here with me, through acknowledging her being *with* as incarnate being, freedom qua freedom, one who is present with me, in whose presence I am. One understands the other person through recognizing him *as* incarnate being, being *with.* Marcel suggests that being recognized by another person "is needed to integrate the self."[180] I seek recognition by others so as to better understand who I am. A purpose of communication is uncovered: to be understood, as a person, by others and by myself. I can express the message I am thinking: "I love you" or "I am sorry" and when the other understands *me*—as one who communicate these messages—I feel recognized as I am. My words and presence communicate what I really mean to say. In communion of presence, we understand the message and the person who is communicating the message. Communion of presence announces co-understanding of participating communicators. Co-understanding constitutes a vital *telos* of communication.

Being able to use our words to express what we think and feel empowers us as communicators. Many of our daily interactions engage these verbal exchanges in ways that provide for successful communicative encounters. But there are moments in which words fail to hold all the meaning I wish to communicate. When the other person recounts her feelings of anxiety or sadness, I have the experience that no words can cross the chasm between us to be of comfort to the other person. In these moments in which verbal communication cannot bridge our gaps, we turn to nonverbal expression, being there, being *with,* being present to. In these encounters, as incarnate being, I communicate being present, being *with, co-esse.* As incarnate being, I manifest presence and being able as body to communicate being *with,* I am communicable. I want to communicate to the other, that she is not alone: this is my *telos* for communicating. I am communicat*ed* to the other as invocation upon her. In communicating presence I am communicating *that I am with* the other.

Standing on opposite sides of a ravine between us, we experience being alone on one side of the incommunicable. Communicating *that* I am *with* her leads to a moment of realization, "you too?"[181] and a "reciprocal openness"[182] opens a between where we are here. No longer alone in my own thoughts and feelings about this situation, we are facing this situation together. Sensing that others are *with* us. Marcel encounters at the nexus of human experience, "we are." I hear echoes of this "we are" in my desire to not be alone. This desire for being is felt as ontological exigence at the heart of my experience. My *telos*—as longings for—is inherent to my very being. Marcel describes these ontological experiences at the heart of existence as an "I long to be *with*" and as an always already "we are." In communion of presence, we awaken to understanding *that* we are not alone. We awaken to a *telos,* or inherent dimension, of communication: the realization that we are always already beings who are *with.* When I participate in communion of presence which is my *telos* as *co-esse* or being *with,* I experience fulfilment at the heart of my being.

Finding a way to cross the breach of misunderstanding, or failed communication, we experience the birth of a new bond between us. We have left behind the viewing the other as a he and moved into a realm where we mutually address the other as a thou. This realm of mutual address indicates the presence of a bond between us, a feeling *for* the other. Communion of presence means a belonging in the region between us. Awakening to being *with* the other person means realizing that I belong *in our being with.* We communicate so as to feel heard and understood, to feel that we belong. We communicate so as to experience belonging with. Our communication's *telos* appears as belonging in the region between us, as belonging in communion of presence.

For Marcel, communion of presence bears dimensions that transcend or point beyond physical, material, or bodily experiences. In a communion of presence, a "path leading from dialectic to love has now been opened."[183] One of the richest experiences of communion of presence is love. In *eros* love, we experience desires to "merge with the other in a higher unity."[184] Our love for the other person is "a pledge and a seed of immortality," when we promise to love the other person forever, we desire that this promise last beyond the confines of this life.[185] Communion in love bears a generative, creative dimension in its power to motivate us to create, promise, even lay down our life for the other person.

This generative, creative power of communion in love sometimes even becomes flesh, a person comes into being, as "incarnate reply to the reciprocal appeal between two beings."[186] In love, both participants commune as incarnate beings, as beings *with,* and communicate a *telos* for love: that their love last forever and live on after their deaths. Communion of presence may be experienced between two persons in communication or may transform into a higher unity in love.

Marcel sees in communion of presence openings for a unity that goes beyond even the higher unity of human love. We participate in communion much like participating in the mystery of being, as an experience of a realm that goes beyond our material, mental, or vital world. When two people love one another, they create a "spiritual interconnection" which is experienced as a bond beyond the merely physical connection which happens at a deeper level of being.[187] Marcel interprets the spiritual world as a dimension for living after and/or beyond this temporal and material world. As described in Chapter One, Marcel believed in his own personal experience of this divine presence, "an absolute Thou," that had made an invocation within his life. Marcel interpreted his experiences with spiritual presences throughout his life as encounters with beings of a transcendent realm.[188] Persons may experience communion as a communion in a spiritual dimension.

Marcel identifies this highest form of love, "*Agape*. . . transcends fusion" and goes beyond unity between persons.[189] In *agape*[190] love, the persons "recognizing each other as different, but loving one another in their very difference" create between themselves "the Highest unity."[191] Agape love indicates an unconditioned love, an accepting love of the other as other. Marcel asks, are there relationships of human love that do not include this transcending dimension? Marcel suggests that all human love which promises that love will be immortal cannot be considered "a closed system" because their love points to a fulfillment beyond this time and space. Human love that promises to be faithful in this life and the next, suggest that their love cannot be contained in a material, earthly dimension. Marcel suggests that any human love that promises eternal love bears transcendent dimensions. If two people promise this eternal love, they will only experience satisfaction in communion if their communion is "centred upon an absolute Thou."[192] Marcel presupposes that all human love is only fulfilled in communion with a divine Thou.[193] This project attends to the dimensions of communion that *precede* what Marcel calls "higher" communion (love) and "highest" communion (love of divine presence). A discussion of what Marcel calls "higher" dimensions falls outside the scope of this study of *telos* for human communication.

SUMMARY: SEEKING COMMUNION

Communion of presence as co-understanding, as incarnate being, as being *with*, and as belonging with offers *telos* for our interpersonal communication. This *telos* of communication uncovers openings to being *with* at the heart of being at all in the world. Marcel's characterization of communion of presence proposes ways we can reach across our interpersonal rifts of noncommunication. Communication transforms to communion of presence when participants share more than words, awaken to being *with* and belonging *with*, and thus understand who they are. In communication, we set out to experience communion of presence, a realm where I am, we are, we are *with*, and we belong *with*.

If communion of presence characterizes the telos of communication, how do we reach this communion? This chapter outlined how Marcel's call to be *with*, to choose an address, and to assume an attitude of *disponibilité* offer ways to communion of presence. Engaging secondary reflection, we remain alert to hearing the many invocations intersubjectivity makes upon us. I hear the sound of being *with* from my birth into existence and my growing sense of self. I am called to recognize that I am *with* others. Each of us has the option to ignore the invocation the other's presence makes upon us; and each of us also has the option to acknowledge the presence of others and to awaken to being *with* the other. Knowing that I am *with* an other person, I choose how I will address them as "he" or as "thou." If I select to address him as "thou," I acknowledge he is a free person that cannot be reduced to one who is less than completely free. Any evaluation I attempt to make that reduces him to anything less than one who is free is a betrayal of who he is, as irreducibly "thou." While these mental acknowledgments describe ways of thinking about the other person, *disponibilité* and *indisponibilité* offer attitudes or ways of being *for* or withheld from the other person. Being *indisponible*, one is absorbed by self and cannot or will not awaken to the invocations made by the other. The *disponible* one is able and open to being *for* the other person. *Disponibilité* positions one to participate in a communion of being *with* and opens one to communion of presence. Awakening to the presence of the other person who is *with* us, positions us to being *with* and *for* the other person, and opens us to more than communication, to being in communion with.

A communion of presence sounds as a far-off ideal realm, a utopia that will be ever elusive for those of us who identify more closely with *indisponibilité* than with *disponibilité*. Marcel always spoke about grounding philosophy in reality, where we are working out the living of being *with* in our actual lives. Marcel emphasizes that his approach to being is always focused on "the return to one's neighbor," because if "we estrange ourselves from our neighbor . . . we can no longer even distinguish being from non-being."[194] If we live in abstract realm of our technologies or our theoretical studies of being, we risk losing the vitally important acknowledgments of human beings who are irreducibly free, communicated and communicating presence, that make invocations upon us. The next chapter turns to reflect on our broken dialogues with our actual neighbors amid our technologies, is communion of presence a real possibility?

NOTES

1. Waterworth, Eva L., and John A. Waterworth, "Focus, Locus, and Sensus: The Three Dimensions of Virtual Experience," *CyberPsychology & Behavior* vol. 4, no. 2 (2001), 203–213, 203.

2. Kwan Min Lee, "Presence, Explicated" *Communication Theory*, vol. 14, no. 1 (2004), 27–50. When one senses physical, material beings, one becomes aware of physical presence. One's experience of being with or before oneself is an experience of self-presence. Sensing that an other thinking actor is *there* with oneself, one experiences social presence. John Short, Ederyn Williams, and Bruce Christie, *Social Psychology of Telecommunications* (London: John Wiley, 1976), 65. Short et al. describe this awareness as salience of the other and salience of the "interpersonal interaction." Schubert, Thomas W., Frank Friedmann, and Holger Regenbrecht, "The Experience of Presence: Factor analytic insights," *Presence* vol. 10, no. 3 (2001), 266-281, 279. Schubert et al. name perception, affect, and interpretation as ways persons cognitively identify the presence of an other.

3. Giuseppe Riva and John A. Waterworth, "Presence and the Self: A Cognitive Neuroscience approach," *Presence-Connect*, vol. 3, no. 3 (2003). Presence as "the feeling of what happens when your being is modified by acts of apprehending something."

4. Erving Goffman, *Behavior in Public Places* (New York: Free Press, 1966), 17. Goffman described copresence as "face-to-face interaction," a physically close interaction of sensing and perceiving the other who is in the same physical space. Celeste Campos-Castillo and Steven Hitlin. "Copresence: Revisiting a Building Block for Social Interaction Theories," *Sociological Theory*, 31, no. 2 (2013), 168–192, 168. Campos-Castillo and Hitlin expanded upon Goffman's description by suggesting "Copresence as the degree to which one actor perceives mutual entrainment (i.e. synchronization of attention, emotion, and behavior) with another actor."

5. Schubert et al., "The Experience of Presence," 279.

6. Eleanor Stump, "Eternity, simplicity and presence," ed. Gregory T. Doolan *The Science of Being as Being* (Washington DC, Catholic University of America Press, 2012), 243–263.

7. Waterworth and Waterworth, "Focus, Locus, and Sensus," 205.

8. Ibid.

9. Gabriel Marcel, "On the Ontological Mystery," in *The Philosophy of Existentialism*, translated by Manya Harari (New York: Citadel Press Kensington Publishing Corp, 2002), 38.

10. Gabriel Marcel, *The Mystery of Being: Reflection and Mystery*, vol. I, translated by G.S. Fraser (South Bend, IN: St. Augustine's Press, 2001), 205.

11. Gabriel Marcel, *Awakenings*, trans. Peter S. Rogers (Wisconsin: Marquette University Press, 2002), 72. Marcel noted how his university studies revealed his idealist leanings understandings of the world. While a student at the *Sorbonne*, he also read American Idealists William Ernest Hocking and Josiah Royce and adopted their focus on intersubjectivity. David W. Rodick, *Gabriel Marcel and American Philosophy: The Religious Dimension of Experience* (Lanham, MD: Lexington Books, 2017). David W. Rodick contended that Marcel's reading of Royce and Hocking shaped Marcel's blended idealist-pragmatic approach. Rodick dedicates a complete study of these in-

fluences. Denis P. Moran, *Gabriel Marcel: Existentialist Philosopher, Dramatist, Educator*. (Lanham, MD: University Press of America, 1992). Moran also summarizes the roles played by American pragmatists Rosiah Royce and William Ernest Hocking in Marcel's works. Royce proposed approaching philosophical problems about divine Being by reflecting on one's own inner experience. Both Hocking and Royce had attempted to reconcile interpersonal relationships within an idealist approach.

12. Gabriel Marcel, *Creative Fidelity*, translated by Robert Rosthal (New York: Farrar, Strauss, and Company, 1976), 31. Gabriel Marcel, "An Autobiographical Essay," in *The Philosophy of Gabriel Marcel*, eds. Paul Arthur Schilpp and Lewis Edwin Hahn (Open Court, 1984), 20. Marcel also describes this experience of being in the office of missing persons, being contacted by persons he felt no connection with initially, but then who shared their anguish at not knowing where their loved ones were. As they talked in person, Marcel frequently made their anguish his own.

13. Gabriel Marcel, *Awakenings*, trans. Peter S. Rogers (Wisconsin: Marquette University Press, 2002), 93.

14. Marcel, "An Autobiographical Essay," in *The Philosophy of Gabriel Marcel*, vol. 17. edited by P.A. Schilpp and L.E. Hahn (Chicago: Open Court, 1984), 20.

15. Rodick, *Gabriel Marcel and American Philosophy*, 1. Rodick describes Marcel's post-war movement as a leaving behind his study of idealism because this approach ignores the "air of the real."

16. Julián Marías, "Love in Marcel and Ortega," in *The Philosophy of Gabriel Marcel*, vol. 17, edited by P.A. Schilpp and L.E. Hahn (Chicago: Open Court, 1984). Henri Bugbee, "A Point of Co-articulation in the Life and the Thought of Gabriel Marcel," *Philosophy Today*, Vol. 19, no. 1 (1975), 61–67. These scholars mention their personal relationship with Gabriel Marcel.

17. Marcel, *The Mystery of Being*, vol. I, 180. Marcel used the French word "avec" which is translated as "with" into English. However, Marcel italicized the word "*avec*" when describing relations between persons. Marcel, "On the Ontological Mystery," 39. He did not use italicized *avec* in his writings to mean "together" or "alongside"; those spatial relations were not comprehensive enough to indicate his idea.

18. Marcel, *Creative Fidelity*, 17. Marcel calls this sense that body is other as "Borrowed otherness."

19. Marcel, *Mystery of Being*, vol. I, 77.

20. Walter Ong, *The Presence of the Word: Some Prolegomena for Cultural and Religious History* (New Haven: Yale University Press, 1967); Thomas J. Farrell, *Walter Ong's Contributions to Cultural Studies: The Phenomenology of the Word and I-Thou Communication* (Creskill, NJ: Hampton Press Inc., 2000). In *The Presence of the Word*, Walter J. Ong offers an in-depth exploration of experiences of presence at the heart of relation to world, other persons, and with divine realm or beings. Ong roots his discussion of presence in communication, as communication involves embodied being, consciousness of the other, and voice as signaling of presence. Ong's work *The Presence of the Word* connects presence, consciousness, voice, and sound with all relation to other persons. Farrell's study of Walter Ong's work provides a significant overview of Ong's major contributions to our study of communication (Farrell,) As noted in Chapter Two, Ong's linking awareness of presence to reflectiveness and communication sounds harmoniously alongside Gabriel Marcel's insights into presence and intersubjectivity.

21. Marcel, *Creative Fidelity*, 147.

22. Julián Marías, "Love in Marcel and Ortega," 555. Although Marcel did not extensively write on communication, Julián Marías remembers when Marcel and Marías attended Heidegger's meeting, "What is this Philosophy?, Marcel frequently called for more discussion about linking philosophy to communicating.

23. Marcel philosophy has been explored for his contributions to understanding intersubjectivity. Kenneth T. Gallagher, *The Philosophy of Gabriel Marcel* (Fordham University Press, 1962), 96–115. Gallagher explores communion and intersubjectivity in a chapter titled "Drama of Communion." Joe McCown, *Availability: Gabriel Marcel and*

the Phenomenology of Human Openness (Missoula, MT: Scholars Press, 1978). Rodick, *Gabriel Marcel and American Philosophy*. John E. Smith, "The Individual, the Collective, and the Community," in *The Philosophy of Gabriel Marcel*, edited by Paul Arthur Schilpp and Lewis Edwin Hahn (Open Court, 1984). Brian Treanor, *Aspects of Alterity: Levinas, Marcel and the Contemporary Debate* (New York: Fordham University Press, 2006).

24. Marcel, *Mystery of Being*, vol. I, 178.

25. Gabriel Marcel, *Metaphysical Journal*, trans. Bernard Wall (Chicago, IL: Henry Regnery Company, 1952), 170. In *Metaphysical Journal*, Marcel first describes how the presence of the other acts as an invocation, later on this chapter this is explained. *Mystery of Being*, vol. I, 179. In this passage from *Mystery of Being*, vol. 1, Marcel also links "invocation" to intersubjectivity when the other person's presence calls upon a person and invokes one into an encounter

26. Marcel, *Metaphysical Journal*, 137. And see comment in *Mystery of Being*, vol. I, 125. Chapter 1, "being in a situation" to describe how a person participates in the world A person experiences being-in-the-world as simultaneously happening to oneself and as an event in which one actively participates.

27. Marcel, *Creative Fidelity*, 29.

28. Gabriel Marcel, *The Mystery of Being: Faith and Reality*, vol. II, translated by René Hague (Chicago: Henry Regnery Company, 1951), 8.

29. Gabriel Marcel, *Homo Viator: Introduction to a Metaphysic of Hope*, translated by Emma Crawford (New York: Harper & Row, 1962), 70.

30. Marcel, *Metaphysical Journal*, 145. As early as 1914, Marcel reflected in his diary, "I am only a given person for myself through the mediating idea of the other for whom I am a given person."

31. Marcel describes self-consciousness, ego, and relation in depth in a chapter, "The Ego and its Relation to Others" *Homo Viator*. Treanor, *Aspects of Alterity*, 69–70. Treanor offers an excellent description of Marcel's use of the terms self, person, and ego.

32. Marcel, *Homo Viator*, 15.

33. Gabriel Marcel, *Being and Having: An Existentialist Diary*, translated by Katherine Farrer (New York: Harper and Row, 1965), 153.

34. Treanor, *Aspects of Alterity*, 69–70.

35. Ibid., 69.

36. Marcel, *Mystery of Being*, vol. II, 9.

37. Marcel, *Mystery of Being*, vol. I, 182.

38. Marcel, *Mystery of Being*, vol. II, 9–10.

39. Mark Wrathall, "Introduction: Metaphysics and Onto-theology," *Religion after Metaphysics*, edited by Mark A. Wrathall (Cambridge: Cambridge University Press, 2003), 2–3.

40. Marcel, *Homo Viator*, 138. In his later work *Homo Viator*, Marcel identifies this as an unpublished note from *Metaphysical Journal*.

41. Dwayne A. Tunstall, *Doing Philosophy Personally: Thinking about Metaphysics, Theism, and Antiblack Racism* (New York: Fordham University Press, 2013), 6.

42. Paul Ricœur, "Gabriel Marcel and Phenomenology," in *The Philosophy of Gabriel Marcel, Library of Living Philosophers, Volume XVII*, edited by P.E. Schillp and L.E.Hahn (Open Court Publishing, 1984). Edmund Husserl, *Cartesian Meditations: An Introduction to Phenomenology*. Translated By Dorion Cairns. Belgium: Martinus Nijhoff, 1970. In the Fifth Meditation, Husserl explores if/how we can know, apperceive, and communicate with other subjects. Husserl offered these *Cartesian Meditations* at the *L'Université de Sorbonne* in 1929 and published these in 1931.

43. Ricœur, "Gabriel Marcel and Phenomenology."

44. Ibid., 483.

45. Marcel, *Mystery of Being*, vol. II, 10.

46. Ibid., 10–11. Marcel expands on his description of this claim as a "given." "The intersubjective nexus" is experienced not as an asserted proposition but as a screen

upon which that-which-is-given appears. Jean-Luc Marion, *Being Given: A Phenomenology of Givenness,* translated by Jeffry L. Kosky (Stanford: Stanford University Press, 2002). Givenness is a richly developed theme within phenomenology. Jean-Luc Marion explores this theme in full at a much later date than these individual Marcellian phrases.

47. Ibid., 17.
48. Ibid., 7, 10.
49. Marcel, "Gabriel Marcel and Phenomenology," 484.
50. Marcel, *Mystery of Being,* vol. II, 10.
51. Marcel, *Homo Viator,* 15. Marcel's full quote is "presence denotes something rather different and more comprehensive than the fact of just being there."
52. Ibid.
53. Marcel, *Mystery of Being,* vol. I, 208. As explored in Chapter Three "At the Heart of Our Restlessness," one uses secondary reflection to study mysteries like being or presence, which "lies beyond the grasp of any prehension," secondary reflection involves awareness that a "purely analytical" reflection will not bring one to understand some phenomena. In secondary reflection, one reflects recursively upon situations in which one is implicated.
54. Tunstall, *Doing Philosophy Personally,* 39. Tunstall makes this distinction between perceiving and appreciating being. Gabriel Marcel, *Mystery of Being,* vol. I, 178. Appreciation is connected to Marcel's term recognizing an other as a unique person who is loved.
55. Marcel, "On the Ontological Mystery," 38.
56. Marcel, *Homo Viator,* 15.
57. Marcel, *Mystery of Being,* vol. I, 209.
58. Marcel, *Homo Viator,* 21. Marcel proposes being a person is not the same as having an ego. One who is self-aware bears an ego but lacks the awareness of being a person. One becomes aware of being a person by feeling responsible for one's actions and by being a subject who acts freely.
59. Marcel, *Creative Fidelity,* 44.
60. Marcel, *Metaphysical Journal,* 137. Marcel's descriptions of being in a situation as the basis for intersubjectivity sound similar to phenomenological proposals of Husserl's "embodied subjectivity" which he offered in 1931, or Heidegger's describing of human experience as being always relating to world and others, *Dasein* as *Mitsein,* in 1927. Marcel's writing on being in a situation from 1918 predates both Husserl and Heidegger.
61. Marcel, "On the Ontological Mystery," 38.
62. Ibid.
63. Richard Zaner, *The Problem of Embodiment: Some Contributions to a Phenomenology of Body* (Belgium: Martinus Nijoff, 1964), 38–39.
64. Marcel, *Mystery of Being,* vol. I, 209.
65. Treanor, *Aspects of Alterity,* 71. Treanor describes this internal-external distinction as Marcel identifying an "ambiguity" in how one experiences the presence of the other. This ambiguous way of experiencing presence reveals presence as a phenomenon that cannot be scientifically explicated or pinned down.
66. Marcel, *Mystery of Being,* vol. I, 208.
67. Henry G. Bugbee Jr., "L'Exigence Ontologique," in *The Philosophy of Gabriel Marcel, Library of Living Philosophers Volume XVII,* edited by P.E.Schillp and L.E.Hahn (Open Court Publishing, 1984) (see note 12), 81–93. Bugbee claims that Marcel's description of the person as participating in being as both an active and receptive participant was "perhaps M. Marcel's most fundamental contribution to the reflective life of the West."
68. Marcel, *Mystery of Being,* vol. I, 181. Marcel differentiates being "beside" an object in distinction from being with a person. One is *with* a person differently than one is with an object.
69. Marcel, *Mystery of Being,* vol. I, 180.

70. Marcel, *Metaphysical Journal*, 170.
71. Ibid.
72. In his introduction to his *Metaphysical Journal,* Marcel suggested that his ongoing reflections—beginning in 1919 in this journal—on the meaning of the word *with* offered significant insight into his approach to intersubjective relations.
73. Marcel, "On the Ontological Mystery," 39.
74. Gabriel Marcel, *The Existential Background of Human Dignity* (Cambridge, MA: Harvard University Press, 1963), 40.
75. Marcel penned this description in 1933. Martin Heidegger had already published Being and Time in 1927, in which he explored the meaning of *Mitsein.*
76. Marcel, "On the Ontological Mystery," 39.
77. Ibid., 38.
78. Marcel, *Being and Having,* 31.
79. Marcel, *Metaphysical Journal*, 247. Marcel states, I feel that my body is "intimately mine."
80. Marcel, "On the Ontological Mystery," 39.
81. Marcel, *Metaphysical Journal,* 168–179. In Marcel's discussion leading up to the following citations from February 24, 1919, in his *Metaphysical Journal,* Marcel explored his experiences of seances during World War I, in which he was trying to ascertain if persons were still alive. Some of his discussions of intersubjectivity begin as his sense of "presence" as phenomena that could be classified as parapsychological or what Marcel calls "metapsychical" experiences like telepathy. Seymour Cain, *Gabriel Marcel* (South Bend, IN: Regnery/Gateway, 1979), 23. Seymour Cain noted that throughout his life, Marcel defended his belief that philosophers should allow for the possibility of "metapsychical" experiences that spring from a real source beyond one's own mind. Rodick, *Gabriel Marcel and American Philosophy,* 22–25. Rodick also addresses Marcel's exploration of metapsychical experiences.
82. Marcel, *Metaphysical Journal,* 170. Marcel wrote this explanation of co-esse in 1919. This predates Heidegger's discussion of *Mitsein* in *Being and Time* which was published in 1927. Martin Heidegger, *Being and Time,* 7th ed., translated by John Macquarrie and Edward Robinson (New York: Harper Collins, 1962).
83. Marcel, *Metaphysical Journal,* 170. This word is italicized because as Marcel noted, "I must use the Latin word" ("On the Ontological Mystery," 39).
84. Marcel, "On the Ontological Mystery," 39.
85. Marcel, *Existential Background,* 41.
86. Marcel, *Mystery of Being,* vol. I, 209.
87. Ibid., 208. I propose this combination of ideas blending "Invoked and evoked" with Marcel's phrase "borrowed otherness" from *Creative Fidelity,* 17.
88. Marcel, *Metaphysical Journal,* 143–146. Marcel frequently used the lower-cased thou for his description of a relation toward the other using a tone of respect. However, Marcel at times used a capitalized "Thou" when referring to relationships between persons. Brian Treanor, *Aspects of Alterity,* 60, 74, 78. When Marcel referred to divine person as thou, he always capitalized, "Absolute Thou." Treanor also uses "Thou" for the other person in intersubjective encounters.
89. Marcel, *Metaphysical Journal,* 143–146; Marcel described a relationship to thou in a 1918 entry of his *Metaphysical Journal,* in 1935 *Being and Having,* in 1940 *Creative Fidelity*, and in 1951 *Mystery of Being: Reflection and Mystery*. See Gabriel Marcel, "I and Thou," translated by Nathaniel Lawrence, in *The Philosophy of Martin Buber, Library of Living Philosophers,* Vol. 12., edited by P.A. Schilpp (Chicago: Open Court, 1967), 41. In Marcel's 1967 reflection on Martin Buber's book *I and Thou,* first published in 1923, Marcel noted a "striking coincidence" that at the same time that Buber wrote his book, Marcel was also describing his own understanding of the human interrelations as oriented toward the other person as thou. Marcel suggested that he and Buber offered reflections on the I-thou relation within a shared historical milieu focused on defining and explicating objectifying relations like I-It. Marcel reflected that philosophers like Ferdinand Ebner, Franz Rosenzweig, or Buber responded similarly to this shared

milieu, considering the absence of or call for an other way of addressing the other person. Although Marcel addressed intersubjective relations as dealing with a thou, he asserted that Buber provided a more thorough exposition of these relations.

90. Marcel, *Metaphysical Journal*, 146.
91. Marcel, *Mystery of Being*, vol. I, 177.
92. Marcel, *Creative Fidelity*, 32, 71.
93. Marcel, *Being and Having*, 106-7.
94. Marcel, *Metaphysical Journal*, 138.
95. Marcel, *Being and Having*, 107.
96. Ibid.
97. Ibid.
98. Ibid., 106–107.
99. Ibid., 107.
100. Marcel, *Metaphysical Journal*, 146.
101. Ibid.
102. Ibid., 147.
103. Marcel, *Creative Fidelity*, 33.
104. Ibid.
105. Otto Friedrich Bollnow, "Marcel's Concept of Availability," in *The Philosophy of Gabriel Marcel*, vol. 17, edited by P. A. Schilpp and L.E. Hahn (Chicago: Open Court, 1984).
106. The translator for *Being and Having: An Existentialist Diary*, Katherine Farrar, used the words "disposability/ nondisposability" for the words *disponibilité / indisponibilité*.
107. Marcel, *Mystery of Being*, vol. I, 163. Marcel, bilingual in French and English, suggested that the English term availability did not fully describe what he meant by the philosophical phrase *disponibilité*, In his lectures in English, he would frequently use the French word *disponibilité* and describe what he meant by the phrase.
108. Kenneth T. Gallagher, *The Philosophy of Gabriel Marcel*. Rodick, *Gabriel Marcel and American Philosophy*. Brendan Sweetman, ed. *A Gabriel Marcel Reader* (Indiana: St. Augustine's Press, 2011). Brian Treanor, *Aspects of Alterity*. Gallagher, Rodick, Sweetman, and Treanor used the untranslated French phrase *disponibilité*. Gallagher sometimes used the English word "disponibility."
109. Ricœur, "Gabriel Marcel and Phenomenology," 485.
110. Rodick, *Gabriel Marcel and American Philosophy*, 25.
111. Rodick, *Gabriel Marcel and American Philosophy*, 27–29. Rodick offers a description of Marcel's "turn to praxis" in his study of intersubjectivity that was inspired by his reading of Josiah Royce's work. Rodick notes Marcel's "commitment to *praxis* as an ontologically efficacious activity." Rodick notes how Marcel offers an intersubjective praxis of hope. Hope is a praxis because "hope transcends the material realm" and hope emerges "in the making." Rodick, 29. This project extends Rodick's identification of intersubjective praxis in Marcel's concepts and this project proposes that *disponibilité* is also an intersubjective praxis.
112. Aristotle. "Nicomachean Ethics" in *Philosophy of Technology: The Technological Condition: An Anthology*. 2nd ed. Edited by Robert C. Scharff and Val Dusek, 19–24 (Malden, MA: John Wiley & Sons, Inc., 2014), 21–22. My approach to praxis engages Aristotle's description of praxis. Aristotle describes praxis as a way of understanding/ acting in the world in distinction from understanding, the Greek word *nous*, and from wisdom, the Greek word, *sophia*. Aristotle suggest praxis involves intelligence used in deliberating about "human concerns." Praxis is a blending of intelligence with deliberation with actions. One engages praxis when one seeks to "expresses rational calculation in pursuit of the best good for a human being that is achievable in action." This intelligence deals with universals and particulars for action.
113. Marcel, "Existence and Objectivity," translated by Bernard Wall, reprint in *Metaphysical Journal* (Chicago: Henry Regnery Company, 1952), 337.
114. Marcel, *Creative Fidelity*, 88.

115. Ibid.
116. Ibid.
117. Marcel, "On the Ontological Mystery," 40.
118. Ibid., 42.
119. Marcel, *Creative Fidelity*, 54.
120. Marcel, "On the Ontological Mystery," 42-3.
121. Marcel, *Creative Fidelity*, 54.
122. Ibid., 88.
123. Marcel, *Homo Viator*, 16.
124. Marcel, "On the Ontological Mystery," 40.
125. Ibid., 41.
126. Ibid., 40.
127. Sam Keen, *Gabriel Marcel* (Richmond, VA: John Knox Press, 1967), 9–16.
128. Marcel, "On the Ontological Mystery," 39.
129. Ibid., 41.
130. Marcel, *Creative Fidelity*, 36.
131. Marcel, *Mystery of Being*, vol. I, 145.
132. Marcel, *Being and Having*, 184, 226.
133. Pietro Prini qtd. in Joe McCown. *Availability: Gabriel Marcel and the Phenomenology of Human Openness*. (Missoula, MT: Scholars Press, 1978), 4.
134. Marcel, "On the Ontological Mystery," 40.
135. Marcel, *Mystery of Being*, vol. I, 163.
136. Marcel, "On the Ontological Mystery," 40.
137. Marcel, *Metaphysical Journal*, 283. Marcel writes this reflection in a 1922 entry of his *Metaphysical Journal*.
138. Ibid.
139. Bugbee, "*L'Exigence Ontologique*," 84.
140. Ibid.
141. Marcel, *Creative Fidelity*, 27.
142. Ibid., 28.
143. Ibid., 28, 89, 91.
144. Marcel, *Mystery of Being*, vol. II, 118.
145. Ibid., 118–119. Marcel states "to give is not to seduce."
146. Ibid., 120.
147. Frank Macke, "Intrapersonal Communicology," *Atlantic Journal of Communication*, 16 (2008), 122–148, 135. This description of communication as more than just the messages aligns with a communicology description of communication. In a Communicology approach to communication, scholars study communication as *more than* verbalized or digitized messages. Frank Macke notes, that for communication to happen, in "conscious and intentional information transfer, something must be felt." Isaac Catt, "Communicology and human conduct: An essay dedicated to Marx," *Semiotica* 204 (2015), 341–360, Catt describes communicology as a human science of embodied discourse.
148. Ramsey Eric Ramsey, *The Long Path to Nearness: A Contribution to a Corporeal Philosophy of Communication and the Groundwork for an Ethics of Relief* (New York: Humanity Press), 1998, 1–2.
149. This work offers a sliver of philosophical insight from Marcel to a vast field of research on interpersonal or relational communication. Decades of scholarly research in books, handbooks, textbooks, and articles provide developed literatures as important backdrop within which Marcel's ideas may add to what we mean by communication. Rob Anderson et al., *Dialogue: Theorizing Difference in Communication Studies* (Thousand Oaks, CA: Sage, 2004). Ronald C. Arnett and Pat Arneson, *Dialogic Civility in a Cynical Age: Community, Hope, and Interpersonal Relationships* (Albany: SUNY Press, 1999). Paul Watzlawick, Janet Beavin Bavelas, and Don D. Jackson, *Pragmatics of Human Communication: A Study of Interactional Patterns, Pathologies, and Paradoxes* (New York: W.W. Norton and Company, 1962).

150. Marcel, *Mystery of Being,* vol. I, 205.

151. Marcel mentions this in *Metaphysical Journal,* 138. Marcel's discussion sounds similar to Charles S. Peirce's explanations of signs. Peirce studies how the non-linguistic and linguistic dimensions of signs and how we use dialogue to create knowledge. In Marcel's description in this citation, Marcel notes the use of signs for the exchange of information.

152. Claude Shannon and Warren Weaver, *A Mathematical Model of Communication* (University of Illinois Press, 1949), 7. Shannon and Weaver described communication as a process involving transmitting information. The communication system includes an "information source" selecting a message, "the transmitter changes the message into the signal which is actually sent over the communication channel from the transmitter to the receiver."

153. Marcel, *Mystery of Being,* vol. I, 205.

154. Marcel, *Metaphysical Journal,* 163. Joe McCown, *Availability,* 39. McCown noted that perhaps Marcel included in later publication of this work this parenthetical reference to Heidegger's term "Mitsein" which loosely translated means "with-being."

155. Marcel, *Mystery of Being,* vol. I, 205.

156. Marcel, *Metaphysical Journal,* 163.

157. Marcel, *Mystery of Being,* vol. I, 207.

158. Ibid.

159. Marcel, *Mystery of Being,* vol. I, 207. The full quote by Marcel is: "The communion in which presences become manifest to each other, and the transmission of purely objectives messages, do not belong to the same realm of being; or rather . . . all transmission of objective messages takes place, if we may so put it, before we have yet reached the threshold of being.

160. Marcel, *Creative Fidelity,* 33. Marcel, "I and Thou," 42.

161. Marcel, *Creative Fidelity,* 33.

162. Marcel, *Mystery of Being,* vol. I, 205.

163. Gabriel Marcel, *Presence and Immortality.* In *The Participant Perspective: A Gabriel Marcel Reader,* edited by Thomas W. Busch, 245–262 (Lanham, MD: University Press of America, 2002), 256.

164. Gabriel Marcel, *Man Against Mass Society.* Translated by G. S. Fraser (Chicago: Henry Regnery Company, 2008), 200.

165. Marcel, *Creative Fidelity,* 33.

166. Quote from *Creative Fidelity,* 35. Marcel clarifies, in this unity, the other person, "while yet remaining separate, has fused into the living unity he now forms with me." *Creative Fidelity* 33.

167. Marcel, *Creative Fidelity,* 33.

168. Marcel, "I and Thou," 46. Martin Buber calls this realm of intersubjective encounter a "between."

169. Marcel, "I and Thou," 46.

170. Ibid.

171. Marcel, *Mystery of Being,* vol. II, 33.

172. Marcel, *Mystery of Being,* vol. I, 205.

173. Marcel, *Being and Having,* 107. According to Marcel, viewing the other as thou, "other qua other," one stands before the other as irreducibly other. Seeing the other as thou, "I treat him and apprehend him qua freedom."

174. Marcel, *Creative Fidelity,* 33.

175. Claude Shannon and Warren Weaver, *A Mathematical Model of Communication* (University of Illinois Press, 1949).

176. A history of the understanding of communication goes far beyond the scope of this project, I offer only two suggested compilations of various approaches to communication. Gehrke, Pat J., and William M. Keith, eds., *A century of communication studies: The unfinished conversation* (New York: Routledge, 2014). Gehrke and Keith edit a series that offers descriptions of the changing definitions of communication happening within the communication discipline studied and discussed within universities in the Unit-

ed States of America. Briankle G. Chang and Garnet C. Butchart, eds., *Philosophy of Communication* (Cambridge, MA, MIT Press, 2012). This text also offers a compilation of philosophical texts to frame or open up understanding of philosophy of communication.

177. Aristotle, *Physics,* in *Western Philosophy an Anthology,* 2nd edition, edited by John Cottingham (Malden, MA: Blackwell, 1996), 413. Aristotle's definition of *telos* appears in *Physics* as the final cause of a being, the end or goal of something. Aristotle states, "'that for the sake of which' something is done" is a cause, the *telos* of that something. Aristotle, *Metaphysics,* in *Philosophy of Technology: The Technological Condition: An Anthology.* 2nd ed., edited by Robert C. Scharff and Val Dusek (Malden, MA: John Wiley & Sons, Inc., 2014), 23. As well in *Metaphysics,* Aristotle describes *telos* as "The end is the good of that thing." *Telos* as an end or good of that thing may mean the fulfillment of that thing.

178. Kent A. Ono and John M. Sloop, "Commitment to Telos—A Sustained Critical Rhetoric," *Communication Monographs,* vol. 59, no. 1 (March 1992), 48–60.

179. Marcel, *Mystery of Being,* vol. I, 208.

180. Marcel, *Homo Viator,* 15.

181. Marcel, *Creative Fidelity,* 33.

182. Marcel, *Man Against Mass Society,* 200.

183. Marcel, *Creative Fidelity,* 33.

184. Marcel, *Man Against Mass Society,* 164.

185. Marcel, *Homo Viator,* 152.

186. Marcel, *Homo Viator,* 70.

187. Marcel, *Homo Viator,* 49.

188. Marcel, *Metaphysical Journal,* 175, 177. Marcel recounted in his autobiographical accounts, that having lost his mother at a young age, he yearned for and sometimes experienced her presence in some way throughout his childhood. During World War I, Marcel also sought to contact spiritual phenomena through seances, using what Marcel called "metapsychical" communication. As noted earlier, Marcel defended his belief that philosophers should allow for the possibility of experiencing "metapsychical" phenomena. Marcel believed he had sensed some encounter *with* real beings beyond this physical, material realm. In some of these experiences, Marcel interpreted these heard or sensed presences as spiritual beings springing from an unknown source. In some of these encounters, Marcel interpreted this presence as a divine Thou who had made a particular invocation upon Marcel in 1929, when Marcel was 41. *Mystery of Being,* vol. II,186. Marcel suggests that human beings, "in so far as we are not things" belong to a "supra-temporal" dimension. Marcel noted how some people "become aware that we are literally arched over by a living reality; it is certainly incomparably more alive than our own." According to Marcel, this spiritual, transcendent, "supra-temporal" "living reality" makes invocations upon persons navigating this space and time.

189. Marcel, *Man Against Mass Society,* 164.

190. Oxford English Dictionary, "agape, n." The Greek word *agape* is commonly translated into English as a type of love that bears divine dimensions. Christian sources offer their definitions for this type of love. To love with *agape* love is to love selflessly, to love with charity, or to love centered on the other(s). Christian authors sometimes distinguish *agape* love from erotic love or affection.

191. Marcel, *Man Against Mass Society,* 164.

192. Marcel, *Homo Viator,* 152.

193. As in Chapter One, I reference this text by Kearney on the presuppositions of faith made by philosophers. While Marcel believes that all human love is ultimately fulfilled in relationship with a Divine Thou, or Person, those who do not share this belief may offer their own theory that human love can reach complete fulfillment on its own, unconnected to any dimension beyond physical, psychological, emotional, and mental dimensions. This work reports on Marcel's theories for relationships and as such includes his belief in Divine dimensions. Richard Kearney, *Debates in Continen-*

tal philosophy: Conversations with Contemporary Thinkers (New York: Fordham University Press, 2004), 259. Kearney addresses the place of faith in philosophy by suggesting that performing a philosophical hermeneutic study.

194. Marcel, *Man Against Mass Society*, 199.

SIX

Seeking Communion as Healing Dialogue

Gabriel Marcel's Philosophy for Now

We often hear about our deep rifts in communication throughout our nations, communities, and relationships. Any mention of this topic for this project, my study of our struggles to dialogue amid technology, brings immediate and passionate confirmation that this struggle is a definite problem for all of us. We see on national level the rifts between political parties, between news channels, and between communities. Our communities use technology, most commonly today as social media, to create barricades around persons with similar beliefs and goals for daily living and interactions and to block out those who do not agree. Yes, we as communities use our technologies to create our shared messages, ideas, and calls to action, but yes, we also use our technologies to unfollow, to ignore, or to cancel the other who stands in different opinions apart from us.[1] For the healing of these seismic rifts in our communities, I defer to the expert peacemakers and diplomatic architects of dialogue.

This work turns our gaze to much closer rifts in communication happening in our homes, in our offices, on our dates, or in a run-in with an acquaintance. If celebrity Twitter wars and Instagram Story rants fade to the background momentarily, our own misunderstandings and disconnections come into focus in our own foreground. We struggle to communicate with the people alongside whom we live, work, and enjoy life. Why are we so often unable to dialogue with the people at our side? This chapter returns to study *our* communication broken by misunderstandings, uncrossable gaps, and absence, *our* interpersonal communication complicated by the presence of our technology.

Our problems with interpersonal communication are often our problems with dialogue. Our struggles in communication appear as misunderstandings of and disconnections from the others. I do not understand what she wants to say and she does not feel understood in the interaction. Our misunderstandings create rifts between us giving us a sense that the other is disconnected from me. As we stand on opposite sides of chasms in our romantic relationships, families, friendships, professional interactions, and our community encounters, we experience broken communication with our actual neighbors. We lack common ground on which to discuss our ideas. We try to have conversations with the other person but trying to converse along rifts of difference between leads to even more frustrating attempts at communication. We shout across distances, the other has to repeatedly ask, "What are you saying?," we try to say what we just said again, the other responds something back, and we have to ask them, "What are you saying?" Our struggles to converse amid differences reveal our ongoing problems with communication as problems to understand and to be understood by the other. Other times in our attempts to communicate, we get the sense that the other person is not really there *with* us. We may be using words to transmit ideas to the other and still feel alone in the interaction. Our feeling alone amid moments of communicating points to a lack of real connection with the other person.

In Chapters One and Five, we explored in depth a common encounter between persons in which communication between persons breaks down. Gabriel Marcel depicted a scenario from his day (the 1950s) of faltering communication happening amid persons present in the same room with no mention of any technologies present. He states, "One might say that what we have with this person, who is in the room, but somehow not really present to us, is communication without communion: unreal communication, in a word."[2] Real communication sounds as communication with communion. Communication with communion is communication as dialogue. Dialogue emerges when we understand and feel understood by the other person and when we are *with* and feel accompanied by the other person. Dialogue sounds as communion between communicators. What might Marcel's texts from the 1950s say about our communication today? Am I aware of who I am as communicator? Do I understand myself as incarnate being? Am I aware of the other person present with me? Are we communicating without communion? Real communication or genuine dialogue assumes new meaning and conditions in today's technological context.

Technological advances in our communication technologies may promise us improved communication as clearer sound and stronger signals. In many contexts, technology makes improved communication possible.[3] In other settings, technology complicates our interactions or attempts at connecting. As explored in Chapter Two, technology has brought changes for the environments in which we interact and for us as persons who are interacting. This project delves into the question, can we find openings for real communication amid our interactions today? Before we explore this question, we first ask, are our problems with communication related to our increased use of technology? Maybe, yes, or maybe our problems in connecting are our ever-present problems as human communicators who struggle to communicate effectively with others.

I find Marcel's writing on technology, on being, and on intersubjectivity offer places to start understanding our own communication amid technology. In Marcel's work, his reflections on technics are not directly connected to his explorations of intersubjectivity. This project, building upon connections made by other scholars,[4] uncovers rich intersections between Marcel's philosophy of technics and Marcel's insight into intersubjectivity. Marcel's insights into our problems in *really* communicating also help us understand our problems communicating amid technology.

Gabriel Marcel offers generative signposts to where and how we may find dialogue today amid our increasingly technological settings. Marcel invites us to step back and look at the often overlooked dimensions of our lives. Our lives have been altered by the many technological changes in our world. First, this chapter attends to Marcel's study of technology as context in which we attempt to communicate with others and asks, how has technology changed how we communicate *with* others? Amid these changes to our communication, we encounter barriers between us that appear to block our attempts at communion *with* others. Our problems with communicating have become metaproblems. Second, I ask, in seeking a solution for our metaproblems, where do we begin? Building upon explorations in Chapters Three, Four, and Five this chapter identifies in Marcel's works openings for communion amid our attempts to communicate. Finally, this chapter closes this project identifying ways these openings for communion *may* bring healing (for) dialogue. Drawing implications from Chapter Five, this chapter identifies Marcel's concept of communion as the seen, felt, and lived goal, as our *telos* for why we attempt to dialogue amid our technology. This chapter summarizes some exploratory wanderings, in Marcel's style as *homo viator*.[5] Any work on Gabriel Marcel's writings does not end with firm conclusions that end discussion. Marcel's writings resist closure. Gabriel Marcel's thoughts leave us question*ing*, ask*ing*, wonder*ing*, in liv*ing* dialogue: dialogue as ongoing interactions among human communicators that was happening long before we began participat*ing* and will continue long after I have left the conversation.

TECHNOLOGY COMPLICATES OUR COMMUNICATION PROBLEMS

Our technologies have immensely improved our channels for and speed of communication. Although technology often improves our communication, sometimes we experience that communication itself breaks down. Do we pause to identify our problems with communicating? Do we think about all the possible reasons for why we struggle to communicate today? We have been immersed in a society that sees all the ways technologies *help* and *improve* our lives. To question the belief that technologies only improve our living and communication makes the questioner anti-progress, anti-science, and anti-advances. Marcel states that this sidelining of questioners about technology and its impact on human experience is not fair to the questioners.[6] Part of being human is to be able to discern if what one does is improving or diminishing one's life. Those who *question* the narratives that say more technological advances are always better are merely assuming a metatechnological perspective.[7] Marcel calls this metatechnological perspective, a reflective practice of taking a critical look at the "concrete relationship" between technical processes and human beings.[8]

When I talk about communicating amid technologies, I focus this discussion on a few technologies used in interpersonal communication.[9] Our face-to-face conversations may happen in-person in actual[10] shared space and time and/or in virtual online spaces shared synchronously or asynchronously. Our face-to-face conversations in actual shared space and time may happen solely between the participants present there or may also include other participants present virtually, via a technological audio, video, or textual interface.[11] In our face-to-face conversations, we may be just two persons conversing, we may be joined by a third face, a phone or phones within eyesight of any or all communicators. Marcel always brought his reflections to a concrete situation, and so I add this specific context of a common communication experience to set the stage.

Marcel asks, how have our technologies changed our being human, our living, our making sense of our world, and our interactions with others in this world.[12] Are we overlooking some aspect of how we as communicators have been impacted by more technologies? Are we aware of how we as communicators are able to and *actually* interact with other communicators amid technological environments? I return to Chapters Two and Chapter Five to bring Marcel's reflections on technology to bear on how we communicate amid technology. I draw a few implications from Marcel's work about how technology changes *our contexts* for and *how* we communicate. While technology may not be the cause of *all* of our communication problems (the technology *users* seem to cause most of the problems), technology definitely changes our contexts in which we attempt to communicate and therefore impacts *how* we communicate. Technology complicates our ongoing problems with communication.

Marcel proposed that technics bring a disconnection from one's own environment. This disconnection from one's local environment leads to a loss of experiences of "a climate of presence and sympathy."[13] In our talking with others via communication technologies, we communicate with the other in an atextual setting, we have no idea of the context in which the person sends or receives the messages. When I read the text of the other person, I have no sense of their presence, of their nonverbal messages. Knowing less about the other person's situation, I am also less likely to understand why they sent the message. A friend sends a short "yes" or "no" to my long question and because I do not know that that friend is in the middle of a work meeting or someone ringing the doorbell, I am less like to feel *with,* to "sympathize" with her.

In a technological environment, we experience our "rhythm of life" has become accelerated and we feel rushed to produce things especially in our workplaces.[14] Our use today of devices all day and night disrupts our daily rhythms, as we find ourselves sending texts or comments at all times of day and night. Our disconnection from our natural biological circadian rhythms leads to compromised physical recovery from our labor and from our social connections.[15] We are always on, tethered via our devices to our communication with others. Marcel suggests that this acceleration experienced in a technological age is accompanied by a loss of the "sense of the sacred."[16] Marcel's sense of the sacred could mean the sense of sacredness of one's physical and emotional dimensions—who one is outside of one's work. A person has more dimensions than just what she produces at her workplace. If she is continually addressed as a laborer who needs to respond to emails day and night, she—as much more than her labor—is no longer recognized as being sacred. Marcel noted, that caught up in an accelerated pace, a person "ignores more and more systematically his condition as a living being."[17] Feeling perpetually connected to social networks around work and profession, one feels as if one does not have time to attend to *who one is* as more than laborer, as living being who needs time to be able to pause, to think, to live.

One who navigates a technological world may so extensively ignore her own condition as a "living being," that she lose sight of the sacredness of her own life. Along with losing one's "sense of the sacred"[18] in the world, one loses one's sense of one's own sacredness. Uprooted from one's community connections, one becomes a producer, one who is defined by her productivity. A technological standard compares the person to the machine. My rest becomes an absence from work and I feel that even my rest should make more productive. Our watches count our steps and give us assessment of our achievements or our falling short of goals (that we set or that were set for us as goals for healthy productivity or productive health). Marcel suggests that one's condition as living being is being more *than* what she produces and that one must pay attention to one's "condition as a living being," to be able to understand what living being means.[19] Marcel notes that many people report feeling disquietude, restlessness, and alienation.[20] These feelings indicate internal disunion. One's intrapersonal dialogue, as where and how one interprets and makes sense of one's life, speaks of loss, uncertainty, and even alienation experienced at one's inner most core. When one pauses to think about how one is doing, one may be tempted to use technological productivity standards for measurement. We will always underperform our technologies, and so we may be tempted to feel our lives are not quite so sacred.

According to Marcel, our technologies distract us from acting reflectively. In technical settings, people develop an increased dependency on their technics. Writing in the 1930s, Marcel notes, "he is increasingly incapable of controlling his technics, or rather of *controlling his own control*" and soon feels "at the mercy of his technics."[21] I am not sure what technologies Marcel refers to in this context, maybe technics as machines. I could replace "technics" with "cell phone" and Marcel's phrase often applies to our relations to our cell phones. Many of us remain connected online almost all the time.[22] This relation of dependence may affect how we think about our world. People considerably dependent on their "gadgets" can live in a "process of automization," thoughtlessly focused on their material possessions and unaware of their own inner world.[23] Marcel also notes, "the more a man becomes dependent on the gadgets whose smooth functioning assures him a tolerable life at the material level, the more estranged he becomes from an awareness of his inner reality."[24] In automatic motions, we check our phones last thing at night and first thing in the morning. We scroll through updates on our newsfeeds as our first thoughts about the day. Maybe these newsfeeds provide us with daily meditations or ideas for reflection. Not all unreflective actions diminish our "awareness of [our] inner reality" the way Marcel suggests. Marcel's invitation offers this question for self-reflection, do I have control of my control of my technologies? I might respond, no and I am okay with that. Marcel's reflections invite us to attend to our automatic or taken-for-granted practices, to glean understanding about *how* I am living with my technologies.

Marcel's insights point to technology significantly changing our human experience: disconnecting us from our environments, distracting us from acting reflectively, and making us less free to connect to others. Changes to our *contexts* in which we interact bring changes to our communication. We feel our messages are taken out of context, we don't pause to reflect before communicating, and we fear misunderstandings may multiply among our larger and larger audiences. Our technologies complicate our already existing problems with communication. If *without* technologies, we already experience, "that what we have with this person, who is in the room, but somehow not really present to us, is communication without communion: unreal communication, in a word,"[25] *with* technologies, our communication becomes *more complicated.*

In our texted interactions, our conversation exchanges can become a mere exchange of words between. Technologically mediated settings prevent us from always knowing if we are speaking with the actual person we think we are in communication with.[26] Studies continue to uncover that social media interactions involve extra steps of authentication to ensure someone is in fact who they say they are.[27] Chances are when we are speaking via text messages with a friend, we are in fact talking to our friend and not to a bot, and yet communication may still not be optimal. As we transmit messages to one another, we may miss tone or context. Because I cannot see context, the other person may be distracted by something else while also texting with me. Our increased technological visual and audio signals often increase our distractions from clear or genuine communication.[28] Yes, words are shared, messages sent, but I may not be certain if the other person really received the meaning of my message. We exchange "signs or symbols," in the words between, and yet fail to communicate presence.[29] Many times, we need to text more words to assure the other person we have understood *more than* the words: "Yeah, I got it," "No, really, I get it." Even when we are face-to-face, talking with the person across from us at dinner, when a third "face"-as-phone is also on the table, we check in to ensure they are really there with us, "Did you hear what I just said?" If Marcel in the 1950s described problems with distraction between persons who are physically in the same room, then we in the 2020s experience increased possibilities for our technological distractions from this person here with us.

Amid our increasing technological connections, we have increased possibilities for experiencing Marcel's depiction of the interpersonal barrier, "he does not understand *me.*"[30] In feeling unheard or misunderstood, we feel separated from the other person—as if the actual or figurative distance between us increases. Our technologically mediated interactions may feel like "impoverished" communication.[31] With waning communication, communicators often feel more and more alone. Many people report a strong connection with the communication device itself, checking in on news, updates, or messages that appear on the device.[32] One's connection with their device need not be a solitary experience, one may be connecting with a person through one's device. But my connection with the person in my shared physical space may be diminished by presence of our technologies. Even if we are together in a shared physical space or video chat from different spaces, the person *with* me actually or virtually may feel farther away from me than a third communicator whose texts I am reading. Many of us experience Marcel's description "what we have with this person, who is in the room, but somehow not really present to us this person, who is in the room, but somehow not really present to us, is communication without communion."[33] I feel as though we are not in fact sharing the same moment, we are not really *with* each other. The space between us is not a meeting ground for encounter. The space between us leaves me feeling isolated from the other, I feel "absent present"[34] or "alone together."[35]

These feelings of being misunderstood in interactions via technology may escalate into larger fears or anxieties. Marcel invites reflection on what technology sometimes introduces. Yes, technology gives us sense of control over some areas of our lives; but in reality, technology does not eliminate all of our worries.[36] Sometimes our technologies present us with different things to worry about. Our increased technological connections and instant information access allows us to know and communicate more quickly. Yet we may quickly share a message we did not mean to and we now have new worries about being misunderstood. While we can now post our message to a much broader audience, this broadcasting brings more possibilities for misunderstanding about or for negative responses to our messages.[37] Persons, hoping to engage in discussion in social media communities, experience uncertainty any time they post, wondering will someone criticize, belittle, or troll one's message or images? Marcel described these feelings as a "preoccupation with security begins to dominate human life."[38] Security in Marcel's context refers to physical and emotional safety amid industrial changes. But our fears for security today could be fears for reputational, physical, emotional and mental security as one's online messages are suddenly exposed virally.[39] Our technologies have altered the scope and vulnerabilities of the context in which we share our messages. While sharing on social media offers us impressively broad horizons for exposure of our ideas, this worldwide exposure opens up possibilities for increased misunderstandings or even attacks from more people.

Losing our sense of the sacredness of our own lives changes our sense of the value of others' lives. Those of us who are governed by fears of being misunderstood or of being ignored, may become self-absorbed, *indisponible,* having no room for the other within us.[40] Our increased engagement in technologies may lead to increased "abstraction" about the others we interact with.[41] People become abstractions to me when I view them as a name, a handle, a profile, a series of texts in my phone.[42] I may forget that behind the name or messages is an actual person. Because my phone abstracts the communicator to her text, images, or videos, I can select when I acknowledge the other person as being *there*. I can put my phone on silence, block a particular person, and the other person's presence is silenced or kept out of my sight. As well, when with actual people, I can lose focus of being *with* them as I am really *with* others via my device. Our communication technologies offer us a myriad of opportunities to avoid having to be with these people who are trying to communicate with us.

Technology may leave us feeling uprooted from or disconnected from our *ecos*, our environments as dwellings. Technology introduces concerns for productivity and effectiveness to our intrapersonal dialogue. In our time absorbed by our technologies, we get distracted from reflecting at all about ourselves or about our interactions. In our distractions and absorptions, we are no longer aware of who, how, or where we are. In our interpersonal interactions, we sometimes use our words to simply transmit messages and lose sight of our being *with* another person. In this loss of awareness or sense of co-presence, we feel overlooked by or misunderstood by the others around us. Our fears that these misunderstandings might escalate in viral proportions sometimes make us build barriers around ourselves to protect ourselves from the others. For Marcel, technology changes our inner and outer world, presenting new obstacles to being human, to being as bodies, and to being *with* others and thus, presenting new obstacles for communication.

(WHERE) ARE THERE OPENINGS FOR COMMUNION?

Every decade of the twentieth and now twenty-first century brings exponential increases in the development and production of faster and more efficient technologies. Solutions for our communication problems in the realm of the technological continue to emerge. They can't hear you? Improve your sound quality. They can't see you? Increase your data transmission speed. Yet, we still face communication problems in our inner world and in our interpersonal worlds. Why do I feel alone even when I am with others? Are they ignoring me? Why won't they listen to me? Why don't they understand me? Where do we start to find a solution for these communication problems? Can we find common ground between us? Can we perfect our selection of words? Can we build our bridges between us? Are we standing at the edge of interpersonal chasms that simply cannot be spanned? One could offer scientific answers about and technological improvements of our channels and ways of communication. Our problems communicating persist. How do we answer our ongoing questions about our communication? How do we deal with our problems in communicating? Culling big data offers us insights into technological impacts on human communication on the macro-scale. Our objective, scientific approach to studying these problems gives us some explanations for why we experience these problems. These problems run far deeper than solutions gleaned from a thorough data analysis or in-depth interviews. I look to glean understanding from data gathered much closer to home: from experiences of my own inner world and from the problems I experience today in my communication.

Seeking understanding of *why* we struggle to communicate, I explored scholarship on communication, communicators, and contexts for communicating. Initially seeking answers to my questions, I encountered this phrase by Susanne Langer, "a philosophy is characterized more by the *formulation* of its problems than by its solution of them."[43] In reflecting on our many complications for communication amid technology, I was encountering more questions about *why* we struggle to really connect with the people around us. Then I came across these ideas in Gabriel Marcel's writing, "philosophy must bring to light the profound but usually unarticulated uneasiness man experiences in this technocratic, bureaucratic milieu where what is deepest in him is not only ignored but trampled underfoot."[44] Philosophy, in its attending to experience, in its articulating the feelings and problems we face, in its questioning *why* our problems persist, brings to light our problems. Philosophy is how we identify our problems and how we formulate our questions. Philosophy of communication is how we identify our problems with communication and how we formulate our questions about communication. Bringing to light our uneasiness, what is deepest in us, and what is sometimes ignored is itself an important act. I do not promise answers for all our questions nor appeasement of our uneasiness. I am bringing into focus some uneasiness we may experience today, I am re-articulating Gabriel Marcel's words explaining how he interpreted the experiences of his day, and you decide if Marcel's interpretation helps us make sense of our experiences today.

Gabriel Marcel suggests that some problems, like our relationships with and amid technology, cannot be solved with a simple solution. As described in Chapter Three, when I think about problems connected to being human, I am thinking about a mystery, a problem in which I, the investigator, am involved.[45] The problem I am thinking about goes beyond a simple solution, the problem is a "metaproblem" requiring more than objective observation. A problem with communication involves me the communicator and so is also a mystery or a metaproblem. Marcel proposes a particular type of secondary reflection for thinking about metaproblems. Secondary reflection *more* reflection a "reflection upon this reflection (in a reflection 'squared')."[46] This reflection upon reflection involves a detaching from experience in order to be able to understand these situations.[47] Marcel named this detaching from experience, "second reflection," "recollection."[48] In secondary reflection, I think about myself who both asks and is asked questions. In recollection, I turn my thought back upon myself to think about what it means to be or what it means that I am. One of our greatest abilities as human persons is our ability to "withdraw" from ourselves and to think about ourselves thinking about ourselves.[49] Marcel suggests that if one engages secondary reflection to address one's metaproblems, one will find insight.

In facing our metaproblem to be addressed: amid our technology, our interpersonal communication sometimes breaks down, where do we start? First, I attend to myself, the one who is reflecting upon this metaproblem. How do I *feel* about *my* communication amid technology?

Chapters Two, Three, and Four described common feelings felt by people today. People feel distracted and displaced by technologies, that they lack control of their technologies, uncertain about events around them, and feel like life itself is broken.[50] I (and it seems like others share this experience) feel an internal disunity about my communication with people around me, about how I talk to or about myself. I feel distracted by my technologies and I cannot always control my own distraction by these technologies. Marcel names these feelings as uneasiness and disquietude. Uneasiness, lack of peace, reveals intrapersonal dissatisfaction *with* my situation. Marcel invites attending to these feelings that some*thing* feels broken or out of place in our internal world. Turning to my internal world, I begin a dialogue with myself, I ask what is wrong or why do I feel this way? An articulation that something feels broken calls my attention to the place that needs healing. I feel disunity within myself, can I experience internal peace? A longing for internal unity points to a longing for communion within. Our dissatisfactions within open us to seeking fulfilment of our desires, our dissatisfactions urge us to seek fulfilment.

Marcel suggests that if one engages secondary reflection upon one's feelings of internal disquietude, one will discover that at the heart of human experience, we all share a longing, a sense of being as being called upon, or ontological exigence. This sense of being as being called upon gives meaning to why I am here as a communicator, as a human being. Marcel's suggestion *that* our being is a calling upon is a theory about existence. In his philosophy, Marcel does not clarify a *who or what* does the calling.[51] Marcel acknowledges that others may propose different theories about how we exist and about the meaning (or lack thereof) in this existence. He offers *an* interpretation of what one might encounter in reflecting on their own being or life. One might interpret Marcel's describing us as being-as-called-upon as us being called upon by communities we are born into. Marcel describes my birth as reply: "I am the incarnate reply to the reciprocal appeal between two beings."[52] My being in the world is in reply to the beings of others. I find myself, in my birth into existence, in communication with others.

At the heart of our experience, we sense that we are not fully in control of being in the world, and thus we experience being called upon, or ontological exigence. This process of understanding oneself as always in communication *with* is an act of intrapersonal dialogue. Chapter Three offers Marcel's ways for seeking this understanding about oneself. I question myself about what it means that I am and I find myself already in dialogue *with* myself about myself. A longing to understand who I am uncovers a longing to understand who I am *in reply to.* My longing to understand myself reveals my longing to understand my own meaning. My longing to understand sends me in search of answers in the world and people around me. Our desires to understand ourselves open us to communicating.

In this search to understand myself, Marcel calls attention to questions we may have overlooked. He asks, what it mean *that* I am here as this body? Again, engaging secondary reflection upon oneself, we ask, "Who am I—I who question being?"[53] When I pause to think about who I am, I encounter the fact, *that* I am. As Chapter Four explored, I hear myself as "borrowed otherness."[54] I am not the root cause of my own being here. I hear my life makes an appeal upon me, "a deep-rooted interior urge. . . an appeal."[55] I hear my being as something *communicated*. I am an "incarnate reply"[56] and I am incarnate being. I participate in the world as incarnate being, as presence. As incarnate being and as presence I am communicated to others. At the heart of my existence, I am a being in communication *with* others. Even if I feel that I am isolated from others, I can never really ignore being *with* others, I am already in position to others, I am already in position for communication. According to Marcel, in one's search for understanding oneself, one will encounter *that* one is incarnate being, incarnate reply, and as body, already in position for communication. Body is how I communicate and am communicated at all. Incarnate being reveals to us that I am able to be *with*, able to be in communion *with*. Marcel suggests that awakening to incarnate being means awakening to one's existential purpose to be *with* others.

We turn to our struggles amid our technologies to be in communion with the persons next to us. Yes, our technologies have brought us to construct more barriers and walls between us and, yes, our feelings of isolation from the other persons may have created chasms between us. Marcel notes how *indisponible* persons have cordoned themselves off from the others. We may use our technologies to create thicker walls between us and the needs or interactions of the others. As we text or communicate with one person while also distracted by other messages, we engage in communication as transmission. When I do not acknowledge you as presence, I interact with you as mere recipient of my messages, as less than irreducibly free, as less than "sacred." Our rifts between us carry ongoing echoes of failed communication. Sometimes our chasms between persons feel unbreachable.

When our communication has diminished to a mere transmission of message, what could transform this interaction? If I feel alone in my interactions via or in spite of our communication technologies, how might I awaken to being *with* others? When I reach out to a friend and no reply appears or when I sit alone together in the same physical space with someone, we are uncommunicative, distanced by a gap spanning between us. Marcel invites turning secondary reflection upon our experiences along these rifts where we feel that the other person is not present with us. In our feeling disconnected from, we feel longings for connection. Our longings for connection, to be heard by, to be acknowledged by, to be understood by all move us to find fulfilment in feeling that we are truly *with* the other. Our feelings of absence between move us to seek more than what we have right now and to feel that we are not alone. We seek *more than* communication.

As Chapter Five described, instead of persisting in absence, I could listen for the presence of the other. One steps back to see who is here *with* oneself and what being here *with* means. If we feel absorbed by our communication technologies how might we communicate being *for* the other person? One has to *be able* to be with and for the other. One has to be able to be in position to acknowledge the other person *with* oneself. The *disponible* one is not self-encumbered and thus can be open *for* the other. *Disponibilité* positions one to be *for* the other person, welcoming the other, and open to whatever experiences emerge between them. One who is able to position oneself in openness to the other and to the emergence of a between is able to be *with*. Our technologies may exacerbate our self-encumbrance as *indisponibilité*. I may happily escape a challenging discussion by checking something (or anything) on my phone. One who is *disponibile* has room for the other person attempting to communicate with her, she is able to be *for* the others

My internal world, my body as incarnate being, and being *with* others reveal openings for more than communication, for communion. Maybe I'll be able to hear the faint cries of the other, the invocations they make upon me. Becoming aware of the invocations the presence of the other person makes upon me, I may be able to see that she is as incarnate being, as presence. This awakening may happen through sensing bodily presence via video or audio channels or in person. Bodily presence *is* "mysterious invulnerability"[57] that cannot be revoked. Changing one's way of seeing, one understands the other as presence, as sacred and irrevocable presence. Someone might make their presence felt or my words might reach someone. If one awakens to reality that I am here *with* you, one breaks the silence between, beginning a dialogue in co-presence: I am with, we are *with*, we are *co-esse*, we are more than communicating, we are in communion with. As in Marcel's detailed interpersonal encounters without communication technologies, participants must become aware *that* they are with and that being *with* is reciprocated. In our communication amid technologies, we may need an extra step of focusing even more for this communion emerging in-person.[58]

Recollection is this listening *for* the end of silence. Recollection is an awakening to the presence of.[59] Recollection hears the beginning strains of being as called upon. Recollection recognizes that someone else also seeks something more. Here I am and here is someone else *with* me. The *telos* of communication appears as shared seeking for communion. When one is in communion of presence, one's being is refreshed and one has "reached the threshold of being."[60] Communion is fulfilling communication. I am fully present, incarnate being, here and now, you are fully present, incarnate being, here and now, and we are fully present *with, to,* and *for* each other, copresent, "coesse," in communion with. My/our *telos* of communication, why we communicate, is to be *with* and to be in communion of presence *with.* Communion of presence sounds for communicators as what we are listening *for.*

SEEKING COMMUNION AS HEALING DIALOGUE: GABRIEL MARCEL'S PHILOSOPHY FOR TODAY

We return to the landscape that we dwell in today. "Things fall apart, the centre cannot hold."[61] Our technological channels for communication transmit our (mis)communication. At times, I cannot understand at all what the other person means to say. Chasms of misunderstanding, absence and alienation gape between us. In our technological world marked by broken communication, we may feel stuck on our own side, in opposition from an other who is nearby but at a seemingly uncrossable distance. In these scenarios, communication loses its purpose. *Why* should I reach out if I may be misunderstood by, ignored by, isolated from, left dissatisfied or in conflict with the other? Communication that only brings more problems seems pointless. The *indisponible* one has assumed this attitude and takes up his position away from the other person. Our experiences of broken communication in and by our world are exacerbated by our feeling that these divisions cannot be healed. Rehabilitating our communication can seem like a daunting task. Many of us "refus[e] to reflect and refus[e] to imagine" how our situation could change.[62] Remaining on either side of our chasms of noncommunication, my life is safer, more controlled, and definitely quieter.

Gabriel Marcel refused to accept that our broken world cannot be healed. One can always "refuse to accept an intolerable situation as final."[63] Marcel calls this refusal hope. Hope means standing before our seemingly unsolvable, broken communication and refusing to accept that this is all we are capable of. For Marcel, "hope transcends imagination" and "transcends all laying down of conditions."[64]

Some voices may say, technologies have so altered our interactions, our virtual interpersonal communication cannot be genuine dialogue.[65] These voices name conditions in which our communication could be considered dialogue and communication in any setting other than actual face-to-face in real life encounters is not genuine dialogue. Other scholars find genuine dialogue possible in technological settings.[66] Marcel's method of hope sees past the current situation, does not rule out real dialogue in technological settings. A position of hope involves thinking beyond what situation currently fills our horizon. Marcel proposes,

> Might we not say that hope always implies the superlogical connection between a return (*nostos*) and something completely new (*Kaïnonti*)?

> Following from this it is to be wondered whether preservation or restoration, on the one hand, and revolution or renewal on the other, are not the two movements, the two abstractly dissociated aspects of one and the same unity, which dwells in hope and is beyond the reach of all of our faculties of reasoning or of conceptual formulation. This aspiration can be approximately expressed in the simple but contradictory words: *as before, but differently and better than before.*[67]

Hope bears a creative dimension: hope takes what is there and imagines something new to emerge. In hope for dialogue, we return to our beginnings of being, that I am and how I am in the world as reply. Hope as renewal calls to mind who we have been as a restoration for who we are today. As incarnate being, my body as my history, *"as before,"* may reveal that I have been unable to communicate with others. Hope challenges this past and overrides my reasons why I remain in my individual realm of noncommunication. Hope aspires to *"differently and better than before."* Yes, we face many problems in engaging in more than communication amid technology, but we refuse to accept this "intolerable situation as final," we hope for "differently and better than before." We try to think about "something completely new (*Kaïnon ti*)" for how we communicate amid technology.

Our problems in human communication cannot be fixed as if these problems were a mere mechanical issue. Scientific solutions fall short of solving all of our interpersonal problems. When human relations are broken, we need healing. In his 1953 writings, Marcel suggested that the dehumanizing effects technology has had on human communities calls for remedies of love to be engaged on a personal and interpersonal level.[68] A remedy implies healing. Gabriel Marcel's philosophy is revolutionary because he addresses this dimension of our human suffering in need of healing. The intrapersonal and interpersonal brokenness felt as feelings or lived as alienation wounds most, if not all, people. We are not just facing mental puzzles to be solved with reasoning or logic. We suffer our communication problems as intrapersonal and interpersonal wounds. Gabriel Marcel's invitation to seek communion offers a remedy, a way for healing our broken communication for now. Healing is a process engaged at individual and communal levels and on individual and communal timeframes.

Healing begins with a vision of communication that is healed and healthy. Healed communication means communion and dialogue. We imagined dialogue—breaching the gap between us—could be experienced when certain conditions are met. Some scholars suggested that genuine dialogue could only happen in the absence of technology, but maybe the conditions for genuine dialogue have changed. I, the communicator, am aware of who I am as incarnate being. I communicate *that I am with* the other. Both communicators awaken to understanding *that* we are with each other. We experience communion of presence.

Communion sounds as genuine dialogue. Dialogue is *more than* conversation because communicators participate in something more than a transmission of messages.[69] In moments of being present with and to each other, meaning emerges in the "interchange."[70] Marcel notes the experience of better understanding oneself *in the* experience of communion. Marcel describes awakening to the presence of the other which "reveals me to myself, it makes me more fully myself."[71] Genuine dialogue points to communicators who share "a ground or position from which one encounters and listens to the Other."[72] Communication scholar Michael J. Hyde notes, we find that our "existence has the structure of a dialogue," as my existence, being, and being *with* are the way I am at all in the world.[73] In dialogue, one is open to and cooperating with the other. In dialogue, we are really *with* the other person and we understand the other person who is *with* us. Dialogue "involves a cooperative, communicative relationship"[74] and mutual attitudes toward the other.[75] Marcel calls this attitude a position or attitude of "*coesse.*"[76] This active communion with the other person may grow more and more intimate. Communion may bear a dimension of love made of the most intimate interpersonal bonds. Marcel calls this bonded love, "agape" love, when the persons "recognizing each other as different, but loving one another in their very difference" create between themselves "the Highest unity."[77] Communion, as dialogue, bears no limits to the breadth and depth of experience.

"In this age of absolute insecurity we live in, true wisdom lies in setting out, with prudence to be sure, but also with a kind of joyful anticipation, on the paths leading not necessarily beyond time but beyond *our* time."[78] Marcel's philosophy invites thinking *beyond* our time and world. To think *beyond* this world of schismed connections, one needs creativity to imagine healing communication. This project was my attempt at finding a path to start on toward healing dialogue for now. Seeking communion appeared as a goal for the journey. Marcel's meandering explorations pointed to paths that he had found healing. Awakening to being seemed to be a good enough place to start a path toward healing dialogue.

Gabriel Marcel's philosophy helps us understand our communication problems and possibilities. We identify communion of presence as a *telos* as where we will experience fulfilling communication. Marcel invites us to reflect on how we live our present situation. Marcel's philosophy invites awakenings. Awaken to attend to the world I live in. Attend to our dwellings marked by technology and our relationships in these contexts. Attend to who I am as being, as communicator, as technology-user, as in relation to. Attend to who others are, as beings, as communicators, as technology-users, and as present with us. Attend to my attitudes toward others as I am *indisponible* (withheld) from or I am *disponible* (available) toward others. Many things are not working in our communication and communities. We cannot fix all the things. We can start to fix *some*thing.

Where do we start? Listen for openings for communication. Awaken to being. Seek communion. Although these actions *sound* theoretical and thus, impractical, I understand these actions to be the opposite of abstract strategies. I join Marcel in viewing listening, awakening to being, seeking communion as *vital*, existential actions. Maybe our having only focused in our education on physical observations of our world has left our ability to engage metaphysical observation underdeveloped. Just because our ability to reflect on being or think about *telos* for communication is underdeveloped does not mean we should dismiss these actions as valid solutions for our problems. Maybe we have overlooked a deeper or higher method of reflection that would provide an alternative solution for our ongoing communication problems. This project offers Gabriel Marcel's practices of listening, awakening to being and seeking communion as ways to start healing. For now, we listen, attend, and seek, hoping for a sound or sign that we are on the way to healing.

Marcel's reflections communicate his journey of seeking, as a *homo viator,* a traveler passing through, open to experiences along the way. Marcel titled his life story, *En Chemin vers Quel Éveil* ("On the Way to What Awakening"). In awakening to anything, one always finds oneself mid-journey, never having fully arrived at total awakening. Awakening to being is not a ten-step journey to enlightenment. On the way, one asks questions of oneself, of life, of others. The awakening happens in the asking along the way.

As described throughout this project, Marcel did not place his hope in technologies *per se* to help us heal. He hoped in persons to understand their complicated relation with their technologies. His writing suggests that anyone who engaged secondary reflection would awaken to being as ontological exigence, as incarnate being, and as being *with.* This could be considered a serious limitation to Marcel's philosophy. Marcel did not describe in depth what *other* interpretations one might glean from this reflection. What if one reflected on the meaning of one's own life and felt it has no meaning? What if one sought to find something at the heart of existence and sensed that nothing is at the heart of being in the world? What if one tried to listen for the invocations of being, and heard no sound? Marcel's philosophy missing a diversity of perspectives might be a limitation if he had proposed a universal system of philosophy. But Marcel never offered a comprehensive philosophical system for understanding the world. Marcel spoke from and about his experience. And these experiences are infused with *his* ongoing hope in each person's journey to lead that person to an understanding about her world shared with others. Marcel did not express hope in technologies to solve our problems, but over and over, he professed a hope in people, in those who use technologies, to heal their relationships.

Before the uncrossable divides between us and along the canyons of failed communication, we desire communication to close the gap between us. The absence of presence exposes our longing for some*thing*. I may not know what perfect communication feels like, but I do know the communication I am experiencing now is not working. In our searching for communication, we continue to look for a narrowing of the distance for possible points of encounter or for building bridges between. Gabriel Marcel describes the discovery of a bridge between us, "when somebody's presence does really make itself felt, it can refresh my inner being; it reveals me to myself, it makes me more fully myself than I should be if I were not exposed to its impact."[79] Presence overshadows absence. Understanding clarifies uncertainty. Seeking communion, we share co-presence amid difference. These experiences of communion may only last a moment. But momentary refreshment offers powerful motivation for continued seeking. We are always on the way, seeking healing for our communication, seeking bridges between, seeking communion. Hoping in communion to heal our broken communication is an active refusal to lose hope in communication for now.

NOTES

1. Nancy Baym. *Personal Connections in the Digital Age* (Maiden MA: Polity Press, 2015). Baym offers an excellent study of the many ways technology may be used to build or weaken our bonds in communities.
2. Gabriel Marcel, *The Mystery of Being: Reflection and Mystery,* vol. I, translated by G.S. Fraser (South Bend, IN: St. Augustine's Press, 2001), 205.
3. Baym, *Personal Connections.* Baym offers insightful explorations of how this interpersonal connectivity is experienced as increased interactivity and as multimodal encounters. Baym's entire book addresses our interpersonal connections in a digital age and her chapter "Community and Networks" provides particular insights into our technology affording us increased connection.
4. Thomas Anderson, "The Body and Communities in Cyberspace: A Marcellian Analysis," *Ethics and Information Technology,* vol. 2, no. 3 (2000). Dennis Cali, "The Ecology of presence and intersubjectivity in Marcel's philosophy of Gabriel Marcel," *China Media Research,* vol. 11, no. 2 (2015), 109–117. Bernard Gendreau, "Gabriel Marcel's Personalist Ontological Approach to Technology," *The Personalist Forum,* vol. 15, no. 2. *Proceedings of the International Conference of Persons, Santa Fe, NM 1999* (University of Illinois Press, 2003): 229–246. Donald F. Traub, *Toward a Fraternal Society: A Study of Gabriel Marcel's Approach to Being, Technology, and Intersubjectivity* (New York: Peter Lang, 1988).
5. Gabriel Marcel, *Homo Viator: Introduction to a Metaphysic of Hope,* translated by Emma Crawford (New York: Harper & Row, 1962), 7. Marcel uses the phrase *homo viator* as wayfarer or traveller.
6. Gabriel Marcel, "On the Ontological Mystery," in *The Philosophy of Existentialism,* translated by Manya Harari (New York: Citadel Press Kensington Publishing Corp, 2002), 30–31. Marcel finds experiences of despair before the problems in the world. He suggests that this "despair consists in the recognition of the ultimate inefficacy of all technics, joined to the inability or the refusal to change over to a new ground—a ground where all technics are seen to be incompatible with the fundamental nature of being." People have a sense that one "can achieve much as his technics."
7. Gabriel Marcel, *Tragic Wisdom and Beyond,* translated by Stephen Jolin and Peter McCormick, edited by John Wild (Evanston: Northwestern University Press, 1973),

199. A practice that is "Metatechnical or metatechnological . . . involves a critical questioning of technology, of the very notion of technique."

8. Gabriel Marcel, *Man Against Mass Society*, translated by G. S. Fraser (Chicago: Henry Regnery Company, 2008), 62. Marcel, writing in 1951, invited people to "ask ourselves about the concrete relationship that tends to grow up between technical processes on the one hand and human beings on the other; and here things become more complicated."

9. Our most used communication technologies as networked technologies are our cell phones used for access to information on the Internet, for chatting via text, video and/or audio modes. Many people use their communication technologies to connect on social media platforms.

10. Kwan Min Lee, "Presence, Explicated" *Communication Theory*, 14, no. 1 (2004), 37. Ulrike Schultze, "Embodiment and Presence in Virtual Worlds: A Review," *Journal of Information Technology*, volume 25, no. 4 (2010), 446n4. Lee differentiates virtual from actual in describing objects. Lee states that "the term actual simply means that something can potentially be experienced by human sensory systems without using technology." Lee and Schultze do not use "real" as an antonym for virtual, because elements that are virtual may also be real.

11. Technology functions as a face, an interface where I face and am faced by others, where we stand before others' worlds watching and listening to their thoughts, musings, messages; and, where we share or withhold our thoughts, musings, and messages.

12. Marcel, "On the Ontological Mystery," 33–34. Marcel notes that those who make technologies engage these technologies with creativity and thought. "Degradation begins at the point where creativeness falls into self-imitation and self-hypnotism, stiffening and falling back on itself." This falling back upon oneself prevents one from using technologies while reflecting.

13. Gabriel Marcel, *The Decline of Wisdom*, translated by Manya Harari (New York: Philosophical Library, Inc., 1955), 3.

14. Marcel, *Homo Viator*, 80.

15. Elizabeth Murnane et al., "Social (Media) Jet Lag: How Usage of Social Technology Can Modulate and Reflect Circadian Rhythms," *Ubicomp* (2015), 843–854.

16. Marcel, *Man Against Mass Society*, 63.

17. Marcel, *Homo Viator*, 81.

18. Marcel, *Man Against Mass Society*, 63.

19. Ibid.

20. Marcel, "Ontological Mystery" 40.

21. Ibid., 31.

22. Andrew Perrin and Madhu Kumar, "About a Quarter of Americans report going online almost constantly" *Pew Research Reports* (May 14, 2018). http://www.pewresearch.org/fact-tank/2018/03/14/about-a-quarter-of-americans-report-going-online-almost-constantly/. In 2018, 3 in 10 U.S. adults report being online "almost constantly."

23. Marcel, *The Mystery of Being*, 25.

24. *Man Against Mass Society*, 41.

25. Marcel, *The Mystery of Being*, 205.

26. Imperva, "2020 Bad Bot Report." Imperva reports that 37% of all Internet traffic is produced by bots. Bots on social media may produce fake news and produce online interactions that may be seen as interpersonal interactions.

27. Youjung Jun et al., "Perceived Social Presence Reduces Fact-checking," *Proceedings of the National Academy of Sciences*, vol. 114, no. 23 (2017), 5976–5981.

28. Keri Stephens et al., "Reconceptualizing Communication Overload and Building a Theoretical Foundation," *Communication Theory*, vol. 27, no. 3 (2017), 269–289. In our face-to-face communication, many people experience distraction due to or by their technologies.

29. Gabriel Marcel, *Metaphysical Journal*, trans. Bernard Wall (Chicago, IL: Henry Regnery Company, 1952), 138.
30. Marcel, *Mystery of Being*, vol. I, 205.
31. Baym, *Personal Connections*, 58–67. Baym offers a summary of why technologically mediated interactions may feel like "impoverished" communication because we lose social cues or increased antagonism
32. Perrin, Andrew and Madhu Kumar, "About a Quarter of Americans Report Going Online Almost Constantly." In 2018, 3 in 10 U.S. adults report being online "almost constantly." Many people in the U.S. go online multiple times and for long periods of time per day. Janna Q. Anderson and Lee Rainie. *Millennials will suffer and benefit due to their hyperconnected lives.* (2012). *Pew Research Center.* Many people engage in a "hyperconnected relationship" with their technological devices, which may enhance or burden relationships. This project does not offer an evaluation of the cost/benefits of being hyperconnected to one's smartphone, this project simply studies this experience.
33. Marcel, *Mystery of Being*, vol. I, 205.
34. Kenneth J. Gergen, "Cell Phone Technology and the Challenges of Absent Presence," in *Mobile Democracy: Essays on Society, Self and Politics*, ed. V.K. Nyíri (Passengen, 2003), 103–114.
35. Sherry Turkle *Alone Together: Why We Expect More from Technology and Less from Each Other* (New York, NY: Basic Books, 2011).
36. Gabriel Marcel, *Being and Having: An Existentialist Diary*, translated by Katherine Farrer (New York: Harper and Row, 1965), 74.
37. Brooke Auxier, "64% of Americans Say Social Media Has a Mostly Negative Effect on the Way Things Are Going in the U.S. Today," *Pew Research* (October 15, 2020). 71% of social media users felt offended by posts, comments or pictures posted to social media (Statista, "Negative Social media Experiences in the U.S. 2017"). In the U.S. in October 2020, persons reported that social media seems to be contributing negatively to the situation of the country. Respondents connect social media responses of hate, harassment, or conflict to spreading hate or inciting violence. One comment, "People say incendiary, stupid and thoughtless things online with the perception of anonymity that they would never say to someone else in person."
38. Marcel, *Man Against Mass Society*, 44.
39. Justin Patchin "Summary of our Cyberbullying Research (2007–2019)" *Cyberbullying Research Center* (July 2019). People continue to experience bullying online or threatening comments in online discussion spaces. Cyberbullying amid teens continues to affect about 21% of all teens who have been cyberbullied and 13% who admit to cyberbullying others. Michael W. Galbraith and Melanie S. Jones, "Understanding Incivility in Online Teaching." *MPAEA Journal of Adult Education*, vol. 39, no. 2 (2010), 1–10. Participants of all ages in online discussions report feeling attacked or ridiculed when participating in public discussions online.
40. Marcel, "On the Ontological Mystery," 40.
41. Marcel, *Man Against Mass Society*, 1.
42. Thomas Anderson, "The Body and Communities in Cyberspace: A Marcellian Analysis," *Ethics and Information Technology* vol. 2. no. 3 (2000), 156. Anderson describes how Marcel's warnings against "abstraction" may take place in how people relate in virtual communities. Anderson finds virtual communities allow one to avoid "commitments of fidelity to real persons in a real world." In virtual communities, one is able to avoid the "concrete situatedness of embodied subjectivity and intersubjectivity." Anderson considers this abstraction brought on by technology an avoidance of living in actual situations.
43. Susanne K. Langer, *Philosophy in a New Key* (Cambridge, MA: Harvard University Press, 1957), 3–4. Langer offered this insight into philosophy woven through the study of media ecology, culture, and philosophy.
44. Marcel, *Tragic Wisdom and Beyond*, 14.

45. Gabriel Marcel, *Creative Fidelity*, translated by Robert Rosthal (New York: Farrar, Strauss, and Company, 1976), 68.
46. Marcel, *Being and Having*, 118.
47. Marcel, "On the Ontological Mystery," 23.
48. Ibid., 25.
49. Marcel, "On the Ontological Mystery," 24. Marcel wonders that one is able in recollection "of taking up my position—in regard to my life; I withdraw from it in a certain way, but not as the pure subject of cognition, *in this withdrawal I carry with me that which I am and which perhaps life is not.* This brings out the gap between my being and my life. I am not my life; and if I can judge my life…it is only on condition that I encounter myself within recollection.
50. Gabriel Marcel, "The Broken World," in *Gabriel Marcel's Perspectives on The Broken World*, translated and edited by Katherine Rose Hanley, 31–152 (Milwaukee: Marquette University Press, 1998), 46.
51. Marcel, *Mystery of Being*, vol. II, 186. In Marcel's descriptions of his beliefs in a world beyond the physical, natural world, he believed in a divine presence, a Thou, who had invited him, already well into his philosophical work at age 41, to believe in a Divine presence. In another description, Marcel suggests that human beings, "in so far as we are not things" belong to a "supra-temporal" dimension and that some people "become aware that we are literally arched over by a living reality; it is certainly incomparably more alive than our own." According to Marcel, this spiritual, transcendent, "supra-temporal" "living reality" makes invocations upon persons navigating this space and time.
52. Marcel, *Homo Viator*, 70.
53. Marcel, "On the Ontological Mystery," 16.
54. Marcel, *Creative Fidelity*, 17.
55. Marcel, *Mystery of Being*, vol. II, 37.
56. Marcel, *Homo Viator*, 70.
57. Ibid., 217.
58. Paul A. Soukup, "Voice, Electronic Media and Belief: Some Reflections on Walter Ong's Perspective on the Electronically Mediated Word," *Explorations in Media Ecology*, 3, no. 1, 7–15. Soukup offers a similar description of this type of extra step of focusing as more extensive interpretation that is needed in situations of electronic media. Soukup describes an in-depth hermeneutic that tests the seeming presence of the other person to ensure that it is really the person we think it is, we must test the images and sounds to ensure we are not being deceived.
59. Marcel, "Ontological Mystery," 23. Marcel notes, "the act whereby I re-collect myself as a unity; but this hold, this grasp upon myself, is also relaxation and abandon. *Abandon* to . . . *relaxation in the presence of* . . . yet there is no noun for these prepositions to govern. The way stops at the threshold."
60. Marcel, *Mystery of Being*, vol. I, 205, 207.
61. William Butler Yeats, "The Second Coming," in *The Collected Poems of William Butler Yeats*.
62. Marcel, "On the Ontological Mystery," 9; *Mystery of Being*, vol. I, 36.
63. Marcel, *Homo Viator*, 34.
64. Ibid., 45, 46.
65. Some scholars propose that genuine dialogue is a misnomer for communication in these mediated settings. Technological settings in which persons are not sharing physical co-presence nor in the same shared physical space and time introduce barriers for genuine dialogue. Michael J. Bugeja, *Interpersonal divide: The Search for Community in a Technological Age* (New York: Oxford University Press, 2005). Bugeja notes that the growing presence of technologies affect how we attempt to communicate interpersonally. Bugeja identifies how the physical separation and lack of nonverbal communication make the disconnection between persons a great if not uncrossable gap between persons. Bugeja suggests that in-person face-to-face interactions contribute to

strengthening our community relationships, while our electronic communication weakens our face-to-face interpersonal relationships.
Michael L. Kent and Petra Theunissen, "Elegy for Mediated Dialogue: Shiva the Destroyer and Reclaiming Our First Principles," *International Journal of Communication* 10 (2016), 4040-4054. Kent and Theunissen note how dialogic theory emerged in a time when face-to-face conversation happened actually not virtually, the persons had to be physically copresent. Kent and Theunissen suggest that calling the two-way communication that takes place in mediated contexts dialogue can sometimes be a misnomer.

66. Rob Andersen. "Anonymity, Presence, and the Dialogical Self in a Technological Culture" In *The Reach of Dialogue: Confirmation, Voice, and Community*, 99–110, eds. R. Anderson, K. N. Cissna, & R. C. Arnett (Cresskill, NJ: Hampton Press, 1994). Jeremy Langett, "Blogger Engagement Ethics: Dialogic Civility in a Digital Era," *Journal of Mass Media Ethics*, 28, no. 2 (2013), 79–90. Deborah Leiter, and John Dowd, "Textual Expectations (Dis) Embodiment, and Social Presence in CMC" in *Interpersonal Relations and Social Patterns in Communication Technologies: Discourse Norms, Language Structures and Cultural Variables*, edited by Jung-ran Park, 32–47 (Hershey, PA: Information Science Reference/IGI Globral, 2010). Joseph B. Walther, "Computer-Mediated Communication: Impersonal, Interpersonal, and Hyperpersonal Interaction," *Communication Research*, vol. 23, no. 1 (1996), 3–43. Scholars offer varying opinions on whether genuine dialogue in a technologically mediated setting is even possible. Many scholars find that new technologies provide even greater interactivity and therefore ground for better dialogue. Technology offers enriched modes for dialogic communication and enhances the dialogic encounter. One could be more present to the other person via technology and show more involved interest in the other person.

67. Marcel, *Homo Viator*, 67.

68. Marcel, *The Decline of Wisdom*, 18–20.

69. Anderson, Rob, Cissna, Kenneth. N., and Ronald C. Arnett, eds. *The Reach of Dialogue: Confirmation, Voice, and Community* (Cresskill, NJ: Hampton Press, 1994).

70. Ronald C. Arnett, "Toward a Phenomenological Dialogue," *Western Journal of Speech Communication*, vol. 45 (1981), 203. Ronald C. Arnett, *Communication and Community: Implications of Martin Buber's Dialogue* (Southern Illinois University Press, 1986). Arnett extracts this description of dialogue from a comprehensive study of Martin Buber's philosophy of dialogue.

71. Marcel, *Mystery of Being*, vol. I, 205.

72. Ronald C. Arnett, *Dialogic Confession: Bonhoeffer's Rhetoric of Responsibility* (Carbondale, IL: Southern Illinois University Press, 2005), 17.

73. Michael J. Hyde, "The Ontological workings of dialogue and acknowledgment" In R. Anderson, L.A. Baxter, & K.N. Cissna (eds.) *Dialogue: Theorizing Difference in Communication Studies* (57–74), Thousand Oaks, CA: SAGE. Rob Anderson et al., *Dialogue: Theorizing Difference in Communication Studies* (Thousand Oaks, CA: Sage, 2004), 62.

74. Michael L. Kent, and Maureen Taylor, "Building Dialogic Relationships Through the World Wide Web," *Public Relations Review*, vol. 24, no. 3 (1998), 324.

75. Richard L. Johannesen, "The Emerging Concept of Communication as Dialogue," *Quarterly Journal of Speech*, vol. 57, no. 4 (1971).

76. Marcel, "On the Ontological Mystery," 39.

77. Marcel, *Man Against Mass Society*, 164.

78. Marcel, *Tragic Wisdom and Beyond*, 213.

79. Marcel, *Mystery of Being*, vol. I, 205.

Bibliography

Abma, Tineke A., Jennifer C. Greene, Ove Karlsson, Katharine Ryan, Thomas A. Schwandt, and Guy A.M. Widdershoven. "Dialogue on Dialogue." *Evaluation* 7, no. 2 (2001), 164–180.

Anderson, Janna Q. and Lee Rainie. "Millennials will Suffer and Benefit Due to their Hyperconnected Lives." *Pew Research Center* (2012). https://www.pewresearch.org/internet/2012/02/29/millennials-will-benefit-and-suffer-due-to-their-hyperconnected-lives/ (accessed October 25, 2020).

Andersen, Rob. "Anonymity, Presence, and the Dialogical Self in a Technological Culture." In *The Reach of Dialogue: Confirmation, Voice, and Community*, edited by Rob Anderson, Kenneth N. Cissna, and Ronald C. Arnett, 99–110. Cresskill, NJ: Hampton Press, 1994.

Anderson, Rob, Kenneth. N. Cissna, and Ronald C. Arnett, eds. *The Reach of Dialogue: Confirmation, Voice, and Community.* Cresskill, NJ: Hampton Press, 1994.

Anderson, Rob, Leslie A. Baxter, and Kenneth N. Cissna. *Dialogue: Theorizing Difference in Communication Studies.* Thousand Oaks, CA: SAGE, 2004.

Anderson, Thomas. "The Body and Communities in Cyberspace: A Marcellian Analysis." *Ethics and Information Technology* 2, no. 3 (2000), 153–158.

———. *A Commentary on Gabriel Marcel's the Mystery of Being.* Milwaukee Wisconsin: Marquette University Press, 2006.

———. "The Nature of the Human Self According to Gabriel Marcel." *Philosophy Today* 29, no. 4 (1985), 273–283.

———. "Technics and Atheism in Gabriel Marcel." *Bulletin de la Société Américaine de Philosophie de Langue Française* 7, no. 1/2 (1995), 59–68.

Appelbaum, David. "Body-Consciousness: Gabriel Marcel's Debt to Maine de Biran." *Bulletin de la Société Américaine de Philosophie de Langue Française* 5, no. 1 (1993), 46–54.

Aristotle. "*Nicomachean Ethics*" in *Philosophy of Technology: The Technological Condition: An Anthology.* 2nd ed., edited by Robert C. Scharff and Val Dusek, 19–24. Malden, MA: John Wiley & Sons, Inc., 2014.

———. *Metaphysics.* In *Philosophy of Technology: The Technological Condition: An Anthology.* 2nd ed., edited by Robert C. Scharff and Val Dusek, 22–24. Malden, MA: John Wiley & Sons, Inc., 2014.

———. *Physics.* In *Western Philosophy an Anthology,* 2nd edition, edited by John Cottingham, 413–415. Malden, MA: Blackwell, 1996.

Arneson, Pat. Ed. *Perspectives on Philosophy of Communication.* West Lafayette, IN: Purdue University Press, 2007.

Arnett, Ronald C. *Communication and Community: Implications of Martin Buber's Dialogue.* Carbondale, IL: Southern Illinois University Press, 1986.

———. *Dialogic Confession: Bonhoeffer's Rhetoric of Responsibility.* Carbondale, IL: Southern Illinois University Press, 2005.

———."Toward a Phenomenological Dialogue." *Western Journal of Speech Communication* 45 (1981), 201–212.

Arnett, Ronald C. and Pat Arneson. *Dialogic Civility in a Cynical Age: Community, Hope, and Interpersonal Relationships.* Albany: SUNY Press, 1999.

Arnett, Ronald C., Celeste Grayson and Christina McDowell. "Dialogue as an 'Enlarged Communicative Mentality': Review, Assessment, and Ongoing Difference." *Communication Research Trends* 27, no. 3 (2008), 3–25.

Auxier, Brooke. "64% of Americans say Social Media Have a Mostly Negative Effect on the Way Things are Going in the U.S. Today," *Pew Research,* October 15, 2020. https://www.pewresearch.org/fact-tank/2020/10/15/64-of-americans-say-social-media-have-a-mostly-negative-effect-on-the-way-things-are-going-in-the-u-s-today/

Bakewell, Sarah. *At the Existentialist Café: Freedom, Being, and Apricot Cocktails with Jean-Paul Sartre, Simone de Beauvoir, Albert Camus, Martin Heidegger, Maurice Merleau-Ponty and Others.* New York: Other Press, 2016.

Bateson, Gregory. *Steps to an Ecology of Mind.* Chicago, IL: University of Chicago Press, 2000.

Baxter, Leslie A. and Barbara M. Montgomery. *Relating: Dialogue and Dialectics,* New York: Guilford Press, 1996.

Baym, Nancy C. *Personal Connections in the Digital Age.* 2nd ed. Maiden, MA: Polity Press, 2015.

Bollnow, Otto Friedrich. "Marcel's Concept of Availability." In *The Philosophy of Gabriel Marcel. The Library of Living Philosophers, Vol. 17,* edited by Paul A. Schilpp and Lewis E. Hahn, 178–197. La Salle, IL: Open Court, 1984.

Buber, Martin. *Between Man and Man.* Translated by Ronald Gregor Smith. Mansfield Center, MA: Martino Publishing, 2014.

Bugbee Jr., Henri G. "L'Exigence Ontologique." In *The Philosophy of Gabriel Marcel. The Library of Living Philosophers, Vol. 17,* edited by Paul A. Schilpp and Lewis E. Hahn, 81–93. La Salle, IL: Open Court Publishing, 1984.

———. "A Point of Co-articulation in the Life and the Thought of Gabriel Marcel." *Philosophy Today* 19, no. 1 (1975), 61–67.

Bugeja, Michael J. *Interpersonal divide: The Search for Community in a Technological Age.* New York: Oxford University Press, 2005.

Busch, Thomas W. *Circulating Being: From Embodiment to Incorporation Essays on Late Existentialism.* New York: Fordham University Press, 1999.

Butchart, Garnet C. *Embodiment, Relation, Community: A Continental Philosophy of Communication.* University Park, PA: Pennsylvania State University Press, 2020.

Cain, Seymour. *Gabriel Marcel.* South Bend, IN: Regnery/Gateway, 1979.

Cali, Dennis. "The Ecology of Presence and Intersubjectivity in the Philosophy of Gabriel Marcel." *China Media Research* 11, no. 2 (2015), 109–117.

Campos-Castillo, Celeste, and Steven Hitlin, "Copresence: Revisiting a Building Block for Social Interaction Theories," *Sociological Theory,* 31, no. 2 (2013), 168–192.

Catt, Isaac. "Communicology and Human Conduct: An Essay Dedicated to Marx," *Semiotica* 204 (2015), 341–360.

Chang, Briankle G., and Garnet C. Butchart. eds. *Philosophy of Communication.* Cambridge, MA, MIT Press, 2012.

Charlton, Noel G. *Understanding Gregory Bateson: Mind, Beauty, and the Sacred Earth.* Albany, NY: State University of New York Press, 2008.

Comte, Auguste, "*Introduction to Positive Philosophy*" in *Philosophy of Technology: The Technological Condition: An Anthology.* 2nd ed. Edited by Robert C. Scharff and Val Dusek, 54–67. Malden, MA: John Wiley & Sons, Inc., 2014.

Collins, James. "Introduction to the Torchbook Edition." *Being and Having: An Existentialist Diary.* Translated by Katharine Farrer, vi–xvii. New York: Harper and Row, 1965.

Ellul, Jacques. "On the Aims of a Philosophy of Technology" in *Philosophy of Technology: The Technological Condition: An Anthology.* 2nd ed. Edited by Robert C. Scharff and Val Dusek, 205–209. Malden, MA: John Wiley & Sons, Inc., 2014.

———. *The Technological Society.* Translated by John Wilkinson. New York: Vintage books, 1967.

———. "Understanding Our Age," in *Perspectives on Our Age Jacques Ellul Speaks on His Life and Work.* Revised edition. Edited by Willem H. Vanderburg, 25–46. Toronto, ON: House of Anansi Press, 2004.

Engnell, Richard A. "Toward an Ethic of Evocative Language: Contemporary uses of Holocaust-Related Terminology." *Southern Journal of Communication* 66, no. 4 (2001), 312–322.

Farrell, Thomas J. *Walter Ong's Contributions to Cultural Studies: The Phenomenology of the Word and I-Thou Communication.* Creskill, NJ: Hampton Press Inc., 2000.

Friedmann, Georges. "The Social Consequences of Technical Progress." *International Social Science Bulletin: Social implications of Technical Change, UNESCO* 4, no. 2 (1952), 243–262.

Friedman, Maurice S. *Martin Buber: The Life of Dialogue.* London: Forgotten Books, 2015.

Gadamer, Hans-Georg. *Truth and Method.* Translated by Joel Weinsheimer and Donald G. Marshall. New York: Continuum, 2004.

Galbraith, Michael W., and Melanie S. Jones. "Understanding Incivility in Online Teaching." *MPAEA Journal of Adult Education,* vol. 39, no. 2 (2010), 1–10.

Gallagher, Kenneth T. *The Philosophy of Gabriel Marcel.* New York: Fordham University Press, 1962.

Gendreau, Bernard. "Gabriel Marcel's Personalist Ontological Approach to Technology." *The Personalist Forum* 15, no. 2. *Proceedings of the International Conference of Persons, Santa Fe, NM, 1999,* University of Illinois Press (2003), 229–246.

Gehrke, Pat J., and William M. Keith, eds. *A Century of Communication Studies: The Unfinished Conversation.* Routledge, 2014.

Gergen, Kenneth J. "Cell phone technology and the challenges of absent presence," in *Mobile Democracy: Essays on Society, Self and Politics,* ed. V.K. Nyíri, 103–114. Vienna: Passengen, 2003.

Gillian, Garth J. "Embodiment in Marcel and Merleau-Ponty." In *The Philosophy of Gabriel Marcel. The Library of Living Philosophers, Vol. 17,* edited by Paul A. Schilpp and Lewis E. Hahn, 499–522. La Salle, IL: Open Court, 1984.

Gilson, Etienne. *Existentialisme Chrétien.* Librarie Plon, 1947.

Goffman, Erving. *Behavior in Public Places,* New York: Free Press, 1966.

Hanley, Katherine Rose. "Gabriel Marcel and Postmodernism." *Bulletin de la Société Américaine de Philosophie de Langue Française* 7 (1995), 122–147.

Hanley, Katherine Rose, "Translator's Preface." In *Gabriel Marcel's Perspectives on The Broken World,* edited by Katherine Rose Hanley, 13-30. Milwaukee: Marquette University Press, 1998.

Hannon, Jason. ed. *Philosophical Profiles in the Theory of Communication.* New York: Peter Lang, 2012.

Heidegger, Martin. *Being and Time.* 7th ed. Translated by John Macquarrie and Edward Robinson. New York: Harper Collins, 1962.

———. *The Question Concerning Technology and Other Essays.* Translated by William Lovitt. New York: Harper Perennial Modern Thought, 2013.

Hernandez, Jill Graper. *Gabriel Marcel Ethics of Hope: Evil, God and Virtue.* London: Bloomsbury Academic, 2011.

Hewitt, Nicholas. *The Cambridge Companion to Modern French Culture.* Cambridge: Cambridge University Press, 2003.

Hogan, Bernie, and Anabel Quan-Haase. "Persistence and change in social media." *Bulletin of Science, Technology & Society* 30, no. 5 (2010), 309–315.

Husserl, Edmund. *Cartesian Meditations: An Introduction to Phenomenology.* Translated By Dorion Cairns. Belgium: Martinus Nijhoff, 1970.

Hyde, Michael "The Ontological Workings of Dialogue and Acknowledgment." In *Dialogue: Theorizing difference in communication studies,* edited by Rob Anderson, Leslie A. Baxter, and Kenneth N. Cissna, 57–74, Thousand Oaks, CA: SAGE, 2003.

Imperva. "The 2020 Bad Bot Report." https://www.imperva.com/resources/ resource-library/reports/2020-Bad-Bot-Report/.

Jeong, Se-Hoon, and Yoori Hwang. "Does Multitasking Increase or Decrease Persuasion? Effects of Multitasking on Comprehension and Counterarguing." *Journal of Communication* 62, no. 4 (2012), 571–587.

Johannesen, Richard L. "The Emerging Concept of Communication as Dialogue." *Quarterly Journal of Speech* 57, no. 4 (1971), 373–382.

Jun, Youjung, Rachel Meng, and Gita Venkataramani Johar. "Perceived Social Presence Reduces Fact-checking." *Proceedings of the National Academy of Sciences* 114, no. 23 (2017), 5976–5981.

Kearney, Richard. *Debates in Continental philosophy: Conversations with Contemporary Thinkers.* New York: Fordham University Press, 2004.

Keen, Sam. *Gabriel Marcel.* Richmond, VA: John Knox Press, 1967.

Kent, Michael L. and Maureen Taylor. "Building Dialogic Relationships Through the World Wide Web," *Public Relations Review* 24, no. 3 (1998), 321–324.

Kent, Michael L. and Petra Theunissen, "Elegy for Mediated Dialogue: Shiva the Destroyer and Reclaiming Our First Principles," *International Journal of Communication* 10 (2016), 4040–4054.

Kierkegaard, Søren. *Philosophical Fragments*. Translated and edited by Johannes Climacus, Princeton University Press, 1985.

Langer, Susanne K. *Philosophy in a New Key,* 3rd edition. Cambridge, MA: Harvard University Press, 1957.

Langett, Jeremy. "Blogger Engagement Ethics: Dialogic Civility in a Digital Era." *Journal of Mass Media Ethics* 28, no. 2 (2013), 79–90.

Langsdorf, Lenore "Epistemology, Tropology, Hermeneutics, and the 'Essence of Language'" In *Experiences Between Philosophy and Communication,* 169–190, edited by Ramsey Eric Ramsey and David James Miller. Albany, NY: SUNY Press, 2003.

Lanigan, Richard L. *Speaking and Semiology: Maurice Merleau-Ponty's Phenomenological Theory of Existential Communication.* No. 22. Berlin: Walter de Gruyter, 1991.

Lee, Kwan Min. "Presence, Explicated." *Communication Theory* 14, no. 1 (2004), 27–50.

Leiter, Deborah, and John Dowd. "Textual Expectations,(Dis) Embodiment, and Social Presence in CMC." in *Interpersonal Relations and Social Patterns in Communication Technologies: Discourse Norms, Language Structures and Cultural Variables,* edited by Jung-ran Park, 32–47. Hershey, PA: Information Science Reference/IGI Globral, 2010.

Macke, Frank. *The Experience of Human Communication: Body, Flesh, and Relationship.* Madison: Fairleigh Dickinson Press, 2016.

———. "Intrapersonal Communicology." *Atlantic Journal of Communication* 16 (2008), 122–148.

Mangion, Claude. *Philosophical Approaches to Communication.* Chicago: University of Chicago Press, 2011.

Marcel, Gabriel. "Author's Preface to the English Edition." In *Metaphysical Journal.* Translated by Bernard Wall, vii-xiii. Chicago: Henry Regnery Company, 1952.

———. "An Autobiographical Essay." In *The Philosophy of Gabriel Marcel. The Library of Living Philosophers Vol. 17,* edited by Paul A. Schilpp and Lewis E. Hahn, 3–65. La Salle, IL: Open Court, 1984.

———. *Awakenings* [a translation of Gabriel Marcel's autobiography *En chemin, vers quel éveil*]. Translated by Peter S. Rogers. Milwaukee, WI: Marquette University Press, 2002.

———. *Being and Having: An Existentialist Diary*. Translated by Katharine Farrer, New York: Harper and Row, 1965.

———. "The Broken World." In *Gabriel Marcel's Perspectives on The Broken World.* Translated and edited by Katherine Rose Hanley, 31–152. Milwaukee: Marquette University Press, 1998.

———. *Creative Fidelity*. Translated by Robert Rosthal. New York: Farrar, Strauss and Company, 1964.

———. *The Decline of Wisdom.* Translated by Manya Harari. New York: Philosophical Library, Inc., 1955.

———. "Drama of a Soul in Exile." Translated by Rosalind Heywood. In *Three Plays: A Man of God, Ariadne, The Funeral Pyre,* 13–34. New York: Hill and Wang, 1958.

———. "An Essay in Autobiography." In *The Philosophy of Existentialism*. Translated by Manya Harari, 104–128. New York: Citadel Press Kensington Publishing Corp, 2002.

———. "Existence and Objectivity." Translated by Bernard Wall. Reprint in *Metaphysical Journal*, 319–339. Chicago: Henry Regnery Company, 1952.

———. *The Existential Background of Human Dignity*. Cambridge, MA: Harvard University Press, 1963.

———. "Foreword." *The Philosophy of Gabriel Marcel* Kenneth Gallagher, NY: Fordham University Press, 1962, xiii–xvi.

———. *Homo Viator: Introduction to a Metaphysic of Hope*. Translated by Emma Crawford. New York: Harper & Row, 1962.

———. "I and Thou." Translated by Nathaniel Lawrence. In *The Philosophy of Martin Buber, Library of Living Philosophers,* Vol. 12. Edited by Paul A. Schilpp, 41–48. La Salle, IL: Open Court, 1967.

———. *Man Against Mass Society*. Translated by G. S. Fraser, Chicago: Henry Regnery Company, 2008.

———. *Metaphysical Journal*. Translated by Bernard Wall. Chicago: Henry Regnery Company, 1952.

———. *The Mystery of Being: Reflection and Mystery,* Vol. I. Translated by G. S. Fraser. South Bend, IN: St. Augustine's Press, 2001.

———. *The Mystery of Being: Faith and Reality,* Vol. II. Translated by René Hague. Chicago: Henry Regnery Company, 1951.

———. "On the Ontological Mystery." In *The Philosophy of Existentialism*. Translated by Manya Harari, 9–46. New York: Citadel Press Kensington Publishing Corp, 2002.

———. *Philosophical Fragments: 1909–1914*. Translated by Lionel A. Blain. South Bend: The University of Notre Dame Press, 1965.

———. *The Philosophy of Existentialism*. Translated by Manya Harari. New York: Citadel Press Kensington Publishing Corp, 2002.

———. *Presence and Immortality*. In *The Participant Perspective: A Gabriel Marcel Reader,* edited by Thomas W. Busch, 245–262. Lanham, MD: University Press of America, 2002.

———. "The Sacral in the Era of Technology." In *Searchings,* translated and edited by Wolfgan Ruf, 41–54, New York: Newman Press, 1967.

———. "Testimony and Existentialism." In *The Philosophy of Existentialism,* Gabriel Marcel, Translated by Manya Harari, 91–103. New York: Citadel Press Kensington Publishing Corp, 2002.

———. *Tragic Wisdom and Beyond, Publication of the Northwestern University Studies Phenomenology and Existential Philosophy*. Translated by Stephen Jolin and Peter McCormick, edited by John Wild. Evanston: Northwestern University Press, 1973.

Marías, Julián. "Love in Marcel and Ortega." In *The Philosophy of Gabriel Marcel. The Library of Living Philosophers, Vol. 17,* edited by Paul A. Schilpp and Lewis E. Hahn, 553–568. La Salle, IL: Open Court, 1984.

Marion, Jean-Luc. *Being Given: A Phenomenology of Givenness*. Translated by Jeffry L. Kosky Stanford: Stanford University Press, 2002.

Marx, Karl. *"Capital: A Critique of Political Economy"* in *Philosophy of Technology: The Technological Condition: An Anthology*. 2nd ed. Edited by Robert C. Scharff and Val Dusek, 74–76. Malden, MA: John Wiley & Sons, Inc., 2014.

Marx, Karl, and Friedrich Engels. "The German Ideology" in *Philosophy of Technology: The Technological Condition: An Anthology,* 2nd ed. Edited by Robert C. Scharff and Val Dusek, 77–79. Malden, MA: John Wiley & Sons, Inc., 2014.

Mickunas, Algis. "Maurice Merleau-Ponty," in Pat Arneson (ed.) *Perspectives on Philosophy of Communication* (Lafayette, IN: Purdue University Press, 2007), 137–162.

Moran, Denis P. *Gabriel Marcel: Existentialist Philosopher, Dramatist, Educator*. Lanham, MD: University Press of America, 1992.

McCown, Joe. *Availability: Gabriel Marcel and the Phenomenology of Human Openness*. Missoula, MT: Scholars Press, 1978.

McLuhan, Marshall and Eric McLuhan, *Laws of Media.* Toronto, CN: University of Toronto Press, 1988.

Miceli, Vincent P. *Ascent to Being: Gabriel Marcel's Philosophy of Communion.* New York: Desclee Company, 1965.

Mullan, Margaret M., "Gabriel Marcel's Approach to Recognizing presence: Being, Body, and Invocation," *Review of Communication* 18, no. 4 (2018), 319–335.

Mumford, Lewis. *Technics and Civilization.* Chicago, IL: University of Chicago Press, 2010.

———. "The Technological Imperative," in *The Meaning of Technology: Selected Readings from American Source,* edited by Montserrat Ginés, 129–131. Barcelona, Spain: Centre de Publicaciones del Campus Nord, 2003.

———. *"Tool Users vs. Homo Sapiens and the Megamachine"* in *Philosophy of Technology: The Technological Condition: An Anthology.* 2nd ed. Edited by Robert C. Scharff and Val Dusek, 381–388. Malden, MA: John Wiley & Sons, Inc., 2014.

Murnane, Elizabeth, Saeed Abdullah, Mark Matthews, Tanzeem Choudhury, and Geri Gray. "Social (Media) Jet Lag: How Usage of Social Technology Can Modulate and Reflect Circadian Rhythms." *Ubicomp* (2015), 843–854.

Nietzsche, Friedrich. *The Gay Science with a Prelude in German Rhymes.* Translated by Josefine Nauckhoff. Cambridge University Press, 2001.

Ong, Walter J. *Orality and Literacy: The Technologizing of the Word.* New York: Methuen, 1982.

———.*The Presence of the Word: Some Prolegomena for Cultural and Religious History.* New Haven: Yale University Press, 1967.

Ono, Kent A. and John M. Sloop. "Commitment to Telos—A Sustained Critical Rhetoric." *Communication Monographs* 59, no. 1 (March 1992), 48–60.

Oxford English Dictionary. "agape, n." *OED Online.* Oxford University Press. www.oed.com.

———."first hand, n., adj., and adv." *OED Online.* Oxford University Press. http://www.oed.com.

———."mirabilia, n." *OED Online.* Oxford University Press. www.oed.com.

Patchin, Justin. "Summary of Our Cyberbullying Research (2007–2019)." Retrieved from *Cyberbullying Research Center* (July 2019; accessed October 20, 2020). https://cyberbullying.org/summary-of-our-cyberbullying-research.

Perrin, Andrew, and Madhu Kumar. "About a Quarter of Americans Report Going Online Almost Constantly" *Pew Research Reports.* (May 14, 2018). http://www.pewresearch.org/fact-tank/2018/03/14/about-a-quarter-of-americans-report-going-online-almost-constantly/.

Pew Research Reports. "Chapter Five: Conflict, Friendships and Technology." *Pew Research.* (2015). https://www.pewresearch.org/internet/2015/08/06/chapter-5-conflict-friendships-and-technology/.

Pew Research Reports. "Usage and Attitudes Toward Smartphones" *Pew Research Report* (2015). http://www.pewinternet.org/2015/04/01/chapter-two-usage-and-attitudes-toward- smartphones/.

Plato. *Gorgias.* Translated by W.C. Helmbold. Boston: Pearson Education, 1952.

———. *Phaedrus.* Translated by W.C. Helmbold and W. G. Rabinowitz. New York: Macmillian Publishing, 1956.

———. "The *Republic,* VIII" in *Philosophy of Technology: The Technological Condition: An Anthology.* 2nd ed. Edited by Robert C. Scharff and Val Dusek, 9–18. Malden, MA: John Wiley & Sons, Inc., 2014.

Postman, Neil. "The Humanism of Media Ecology." *Proceedings of the Media Ecology Association 1,* 10–16. https://www.media-ecology.org/resources/Documents/Proceedings/ v1/v1-02-Postman.pdf.

———. The Reformed English Curriculum." In *High School 1980: The Shape of the Future in American Secondary Education,* edited by A. C. Eurich, Vintage Books, 1970.

Prini, Pietro. "A Methodology of the Unverifiable." In *The Philosophy of Gabriel Marcel: The Library of Living Philosophers, Vol. 17*, edited by Paul A. Schilpp and Lewis E. Hahn, 205–239. La Salle, IL: Open Court, 1984.

Ramsey, Ramsey Eric. *The Long Path to Nearness: A Contribution to a Corporeal Philosophy of Communication and the Groundwork for an Ethics of Relief.* New York: Humanity Press, 1998.

Randall, John Herman. *The Making of the American Mind.* New York: Columbia University Press, 1976.

Ricœur, Paul. "Gabriel Marcel and Phenomenology." In *The Philosophy of Gabriel Marcel: The Library of Living Philosophers, Vol. 17*, edited by Paul A. Schilpp and Lewis E. Hahn, 471–494. La Salle, IL: Open Court, 1984.

Riva, Giuseppe, and John A. Waterworth. "Presence and the Self: A Cognitive Neuroscience approach," *Presence-Connect* 3, no. 3 (2003). https://www8.informatik.umu.se/~jwworth/Riva-Waterworth.htm.

Rodick, David W. *Gabriel Marcel and American Philosophy: The Religious Dimension of Experience.* Lanham, MD: Lexington Books, 2017.

Rosthal, Robert. "Translator's Introduction" In *Creative Fidelity*, by Gabriel Marcel, ix–xxvi. New York: Farrar, Straus and Giroux Company, 1976.

Scharff, Robert C. and Val Dusek. "Introduction. In *Philosophy of Technology: The Technological Condition: An Anthology.* Second Edition, Edited by Robert C. Scharff and Val Dusek, 3–7. Malden, MA: John Wiley & Sons, Inc. (2014).

Scharff, Robert C., and Val Dusek. *Philosophy of technology: The Technological Condition: An Anthology.* Second Edition, edited by Robert C. Scharff and Val Dusek. Malden, MA: John Wiley & Sons, Inc. (2014).

Schilpp, Paul Arthur, and Lewis Edwin Hahn, editors. *The Philosophy of Gabriel Marcel.* La Salle, IL: Open Court, 1984.

Schloesser, Stephen. *Jazz Age Catholicism: Mystic Modernism in Postwar Paris, 1919–1933.* Toronto: University of Toronto Press, 2005.

Schmitz, Alfred O. "Marcel's Dialectical Method." In *The Philosophy of Gabriel Marcel: The Library of Living Philosophers, Vol. 17*, edited by Paul A. Schilpp and Lewis E. Hahn, 159–75. La Salle, IL: Open Court, 1984.

Schrag, Calvin. *The Self After Postmodernity.* New Haven: Yale University Press, 1997.

Schubert, Thomas W., Frank Friedmann, and Holger Regenbrecht, "The Experience of Presence: Factor Analytic Insights," *Presence* 10, no. 3 (2001), 266–281.

Schultze, Ulrike. "Embodiment and Presence in Virtual Worlds: A Review." *Journal of Information Technology* 25, no. 4 (2010), 434–449.

Segal, Howard P. "Introduction." In *Life in a Technocracy: What it Might be Like,* by Harold Loeb, i–xxxviii. Syracuse, NY: Syracuse University Press, 1996.

Shakespeare, William, *Macbeth,* in Harold Bloom's, *Shakespeare: The Invention of the Human*, New York: Riverhead Books, 1998.

Shannon, Claude E., and Warren Weaver. *A Mathematical Model of Communication.* University of Illinois Press, 1949.

Short, John, Ederyn Williams, and Bruce Christie, *Social Psychology of Telecommunications* London: John Wiley, 1976.

Slegers, Rosa. "Le point sur les i: Concrete and Philosophical approaches to Commitment and Writing in Gabriel Marcel's Work." *Renascence* 55, no. 3 (2003), 259–269.

Smith, John E. "The Individual, the Collective, and the Community." In *The Philosophy of Gabriel Marcel: The Library of Living Philosophers, Vol. 17*, edited by Paul A. Schilpp and Lewis E. Hahn, 337–349. La Salle, IL: Open Court, 1984.

Soukup, Paul A. "Voice, Electronic Media, and Belief: Some Reflections on Walter Ong's Perspective on the Electronically Mediated Word." *Explorations in Media Ecology* 3 no. 1 (2004), 7–15.

Stephens, Keri K., Dron M. Mandhana, JiHye J. Kim, Xiaoqian Li, Elizabeth M. Glowacki, and Ignacio Cruz. "Reconceptualizing Communication Overload and Building a Theoretical Foundation." *Communication Theory* 27, no. 3 (2017), 269–289.

Strate, Lance. *Echoes and Reflections: On Media Ecology as a Field of Study*. Cresskill, NJ: Hampton Press, 2006.
———. *Media Ecology: An Approach to Understanding the Human Condition*. New York: Peter Lang, 2017.
Straus, Edwin W. and Michael Machado. "Marcel's Notion of Incarnate Being." In *The Philosophy of Gabriel Marcel: The Library of Living Philosophers, Vol. 17*, edited by Paul A. Schilpp and Lewis E. Hahn, 123–155. La Salle, IL: Open Court, 1984.
Sweetman, Brendan. "Introduction." *A Gabriel Marcel Reader*, 1–8, edited by Brendan Sweetman. South Bend, IN: St. Augustine's Press, 2011.
———. Editor. *A Gabriel Marcel Reader*. South Bend, IN: St. Augustine's Press, 2011
Treanor, Brian. *Aspects of Alterity: Levinas, Marcel, and the Contemporary Debate*. NY: Fordham University Press, 2006.
Tunstall, Dwayne A. *Doing Philosophy Personally: Thinking about Metaphysics, Theism, and Antiblack Racism*. New York: Fordham University Press, 2013.
Turkle, Sherry. *Alone Together: Why We Expect More from Technology and Less from Each Other*. Basic Books, 2011.
Traub, Donald F. *Toward a Fraternal Society: A Study of Gabriel Marcel's Approach to Being, Technology, and Intersubjectivity*. New York: Peter Lang, 1988.
Vanderburg, Willem H., Ed. *Perspectives on Our Age Jacques Ellul Speaks on His Life and Work*. Toronto, ON: House of Anansi Press, 2004.
Wahl, Jean Andre. *Philosophies of Existence: Introduction to the Basic Thought of Kierkegaard, Heidegger, Jaspers, Marcel and Sartre*. Translated by F.M. Lory. London: Routledge, and Kegan Paul, 1969.
Walther, Joseph B. "Computer-Mediated Communication: Impersonal, Interpersonal, and Hyperpersonal Interaction." *Communication Research* 23, no. 1 (1996), 3–43.
Waterworth, Eva L. and John A. Waterworth. "Focus, Locus, and Sensus: The Three Dimensions of Virtual Experience." *CyberPsychology and Behavior* 4 no. 2 (2001), 203–213.
Watzlawick, Paul, Janet Beavin Bavelas, and Don D. Jackson. *Pragmatics of Human Communication: A Study of Interactional Patterns, Pathologies, and Paradoxes*. New York: W.W. Norton and Company, 1962.
Wrathall, Mark, "Introduction: Metaphysics and Onto-theology," In *Religion after Metaphysics*, edited by Mark A. Wrathall. Cambridge: Cambridge University Press, 2003.
Yeats, William Butler. "The Second Coming." *The Collected Poems of William Butler Yeats*. https://www.poetryfoundation.org/poems/43290/the-second-coming.
Zaner, Richard. *The Problem of Embodiment: Some Contributions to a Phenomenology of the Body*. 2nd ed. Belgium: Martinus Nijoff, 1964.

Index

About the Author

Margaret M. Mullan is assistant professor at East Stroudsburg University of Pennsylvania, teaching classes on public relations, intercultural communication, and dialogue and social media. She completed her doctorate in rhetoric from Duquesne University, with concentrations in interpersonal communication and the rhetoric of technology. Her scholarly research has addressed dialogue and social media, peace and conflict, dialogue in online communities, and listening. She has been published in *Review of Communication*, *Journal of Religion, Media and Digital Culture*, and *Journal of the Association of Communication Administration*.

www.ingramcontent.com/pod-product-compliance
Ingram Content Group UK Ltd.
Pitfield, Milton Keynes, MK11 3LW, UK
UKHW012119150926
475560UK00002B/181

9 781793 621771